THE GREAT
TOWPATH WALK
From London to York

Grand Union Canal at Leighton Buzzard

The Great Towpath Walk

From London to York

BRIAN BEARSHAW

Line illustrations
by David Chesworth

ROBERT HALE · LONDON

Robert Hale Limited
Clerkenwell House
Clerkenwell Green
London EC1R 0HT

British Library Cataloguing in Publication Data

Bearshaw, Brian
The great towpath walk from London to York.
1. Walking – England   2. Canals – England
3. England – Description and travel –
1971-
I. Title
914.2'04858   DA632

ISBN 0-7090-3279-X

The author and publishers wish to thank
Eyre & Spottiswoode for permission to use extracts from
L. T. C. Rolt's book, *Narrow Boat*.

Set by Rowland Phototypesetting Ltd
Printed in Great Britain by
St Edmundsbury Press Ltd, Bury St Edmunds, Suffolk

Contents

	List of Illustrations and Maps	7
	Preface	11
1	Westminster to Brentford	17
2	Brentford to Watford	31
3	Watford to Cheddington	41
4	Cheddington to Great Linford	53
5	Great Linford to Long Buckby	61
6	Long Buckby to Foxton	69
7	Foxton to Birstall	77
8	Birstall to Trentlock	89
9	Trentlock to Gunthorpe	99
10	Gunthorpe to Newark	111
11	Newark to Dunham on Trent	121
12	Dunham on Trent to Gainsborough	133
13	Gainsborough to Althorpe	143
14	Althorpe to Goole	157
15	Goole to Selby	171
16	Selby to York	181
	Bibliography	197
	Index	199

List of Illustrations and Maps

Drawn by David Chesworth

Grand Union Canal at Leighton Buzzard	*frontispiece*
Westminster Abbey . . . the start of the 304-mile walk	19
The Houses of Parliament	20
Statue of Sir Thomas More in front of Chelsea Old Church	25
Hammersmith Bridge	27
The Fisheries Inn on the Grand Union Canal	36
Watford Ironbridge Lock	44
Cottage at the entrance to Berkhamsted Castle	49
Leighton Buzzard with its Market Cross	56
An original Grand Junction Canal Company's milestone	58
Windmill at New Bradwell	64
Five-rise staircase locks at Foxton	72
Where the branch leaves for Market Harborough at Foxton Locks	74
Thirteenth-century church at Newton Harcourt	79
Gargoyles look out over the Grand Union Canal in Leicester	82
Friars Mill at Leicester	84
Bridge over the Grand Union Canal at Syston	91
St James's Church in Normanton on Soar	94
The White House on the River Soar	96
The motorway-type sign near Trentlock	100
Statue of Sir Robert Juckes Clifton, Nottingham	104
Footbridge across the River Trent	107
Mill House at Caythorpe	113
Newark Castle	118
Parish church of St Mary Magdalen, Newark	124
Castle Barge on the River Trent at Newark	127
The church of St Peter in Chains at Laneham	136

Cooling towers of Cottam power station   141
Gainsborough Old Hall   146
Harbour at West Stockwith   150
Fifteenth-century church at Althorpe   153
Blacktoft Sands Nature Reserve, where the Rivers Trent
 and Ouse join   160
Jetty at Whitgift on the River Ouse   166
New Bridge Hotel at Goole   174
Selby Abbey   178
Cottages at Cawood alongside the River Ouse   184
Entrance gate to the Bishop's Palace at Bishopthorpe 187
The Shambles at York   189
The Guild Hall, close to the River Ouse in York   193
Towers of York Minster   195

Maps

London to York   10
Westminster to Brentford   16
Brentford to Watford   30
Watford to Cheddington   40
Cheddington to Great Linford   52
Great Linford to Long Buckby   60
Long Buckby to Foxton   68
Foxton to Birstall   76
Birstall to Trentlock   88
Trentlock to Gunthorpe   98
Gunthorpe to Newark   110
Newark to Dunham on Trent   120
Dunham on Trent to Gainsborough   132
Gainsborough to Althorpe   142
Althorpe to Goole   156
Goole to Selby   170
Selby to York   180

All maps based upon Ordnance Survey maps with the permission of the controller of Her Majesty's Stationery Office. Crown Copyright Reserved.

Again, to my dear, understanding wife, Jeanette

YORK

RIVER
OUSE

●MANCHESTER

●LINCOLN

RIVER
TRENT

NOTTINGHAM●

●LEICESTER

BIRMINGHAM
●

●NORTHAMPTON

Stoke Bruerne ○

GRAND UNION
CANAL

OXFORD●

LONDON

RIVER THAMES

Preface

The thought of doing a long-distance walk by water has interested me for some time. It was when I was researching for *Towpaths of England* that I realized – what I should already have known – that England is linked by its waterways as well as by its road system. After all, that was why so many water routes were created.

The advantages were the same as those which made me walk the canals for the towpaths book. You could not get lost – theoretically anyway; it was flat, easy walking, and you could just about guarantee that no irate farmer or landowner was going to pursue you. A very high proportion of the walk was bound to be in the country – and deep in the heart of the country at that – with all the attraction of the country: the peace, the solitude, the bird and animal life, the true feel of England. The big attractions for me were the peace and the pace of life found alongside water.

I started by the Thames at Westminster, walked 155 miles from Brentford on the Grand Union Canal, which took in the River Soar, covered ninety-three miles of the River Trent and finally walked up the River Ouse for forty-four miles to York. I found the Trent had the greatest appeal, and one day I must walk the other way, from Nottingham to its source. Between London and York I touched only two cities, Nottingham and Leicester, but took in a host of villages, most of which have managed to evade the ever-increasing tentacles of the road system. There were also market towns of great age, such as Leighton Buzzard and Selby, and fine old towns, such as Uxbridge, Berkhamsted, Newark – perhaps the best on the walk – and Gainsborough. Yet there was no denying the attractions of the Thames at the start of the walk. There was more to see here than in any other twelve miles all the way to York.

There was, too, always the unexpected – the grave of the highwayman at Boxmoor, close to where he was hanged, the pub at Uxbridge where Crown and Parliament tried to end the Civil

War. There were halls and castles, nature reserves and trails, ancient churches, modern power stations with cooling towers (which I find unusually attractive), ships and boats and . . . the wonderful quiet.

It is 193 miles from London to York by road. It was 304 by water. Of those 304 I had previously walked about thirty-five, all on the Grand Union. I had never before walked alongside the Trent or the Ouse, and it was here, by the rivers, that I thought I might encounter most problems. I knew I would have little trouble with the Grand Union, and then only where it became overgrown and was difficult to get through, but the Trent and the Ouse, I thought, might not welcome walkers along its banks in some areas. So while I expected the Grand Union, which formed the first half of my journey, to be relatively trouble-free, I was ready for problems on the second half – from Trentlock, below Nottingham, right through to York. As it turned out, the Trent and the Ouse – the Trent in particular – were immensely satisfying and trouble-free and provided everything I expected from a walk by an English river.

I spent sixteen days getting from London to York and found very little accommodation close to the water. It would have been no use walking and hoping to find somewhere to stay overnight. In every case but one I booked ahead and sometimes had to walk a mile or two to find my pub, farm or guest house. I found the best way of discovering good accommodation was through the various Information Centres, and I have listed, with their telephone numbers, those that take in the districts on the walk. The Yellow Pages are useful too, as is *The Ramblers' and Cyclists' Bed and Breakfast Guide*.

I only once walked through a day not knowing where I was to stay, after being assured beforehand that there was a guest house in the village which always had a Bed & Breakfast notice outside. Unfortunately I could not telephone to book ahead, and when I got there they were having alterations made to the house and could not take me. As it was nearly dark and fairly late and I had had one of my longest walking days, I was ready to break down and howl until I came across the canalside shop at Buckby Locks run by Brenda and Philip Walker. Without hesitation Mrs Walker offered to give me a bed for the night . . . and I shall for ever be grateful. But it is too risky to walk without booking accommodation, because so much of this walk is deep in the country, with little human life.

There are plenty of pubs along the way – most provide meals if not bed and breakfast – but there were few waterside cafés.

For those who prefer to deal in kilometres rather than miles I have included those as well at the beginning of each chapter. For people unacquainted with the canal system, they will find that the bridges on the Grand Union, though not on the rivers, are numbered. Although the Grand Union is now an amalgamation of several canals, the numbering from the original independent canals is retained with new starts being made at Buckby and Leicester on the way north from Brentford.

Although I walked from Westminster to York Minster, this was no ecclesiastical exercise. Yet I did explore several churches which I think are still fine examples of village history – almost the only examples untouched by time. I was particularly taken by the church at Holme, near Newark, with the first-floor room where Nan Scott watched her friends and neighbours going to their funerals during the Plague; the one at West Stockwith, which has a ship's carpenter's memorial; another at Laneham, with its original door which has been swinging on the same hinges for nearly 900 years. For those who like their places of worship to be grander, there was Westminster Abbey, Leicester and Nottingham Cathedrals, Selby Abbey and York Minster.

The walking itself was a great joy. Out of 304 miles by water, I was forced to miss only about fifteen. I was able to stay alongside the Grand Union until within two miles of its end at Trentlock below Nottingham, where there was no way across the water. Some miles were lost in Nottingham as industry claimed the land on the banks of the Trent, and I lost more on the last leg of the journey when the Ouse turned north and there was no way by the river. The days when ferries abounded on the Trent, in particular, and on the Ouse are long gone. Bridges are scarce but thankfully the footpath was continuous at least up one side of the Trent. The only real problem came near Newark, where the path ran out at Farndon, but a call to Newark 702155 enabled me to cross the river the old-fashioned way, by ferry, at a modern-day price, 50 pence.

There was only one really poor section of path which was overgrown and exasperating and made walking tiresome, and that was for about fifteen miles of the Grand Union Canal leading up to Foxton Locks in Leicestershire. It is as well to try to hitch a lift on a boat to get past this stretch. For me, it was one of those days . . . no boats came. The first seventy miles of the Grand Union were absolutely perfect, and there was also a fine stretch around Leicester, where the city fathers have taken pride in their waterway. I expected the Trent to be typically riverlike, wandering through fields, flooded and muddy in parts. It did spend much of its time in fields but the floodbanks which have become necessary to keep the river in its place provided magnificent walking and fine views.

I started out not knowing quite what I would find. I finished at York after the most satisfying succession of walks I had undertaken. I was unhappy only with the overgrown part of the Grand Union Canal but found much compensation, particularly on the Trent with its beauty and especially its peace, so far from the madding crowd.

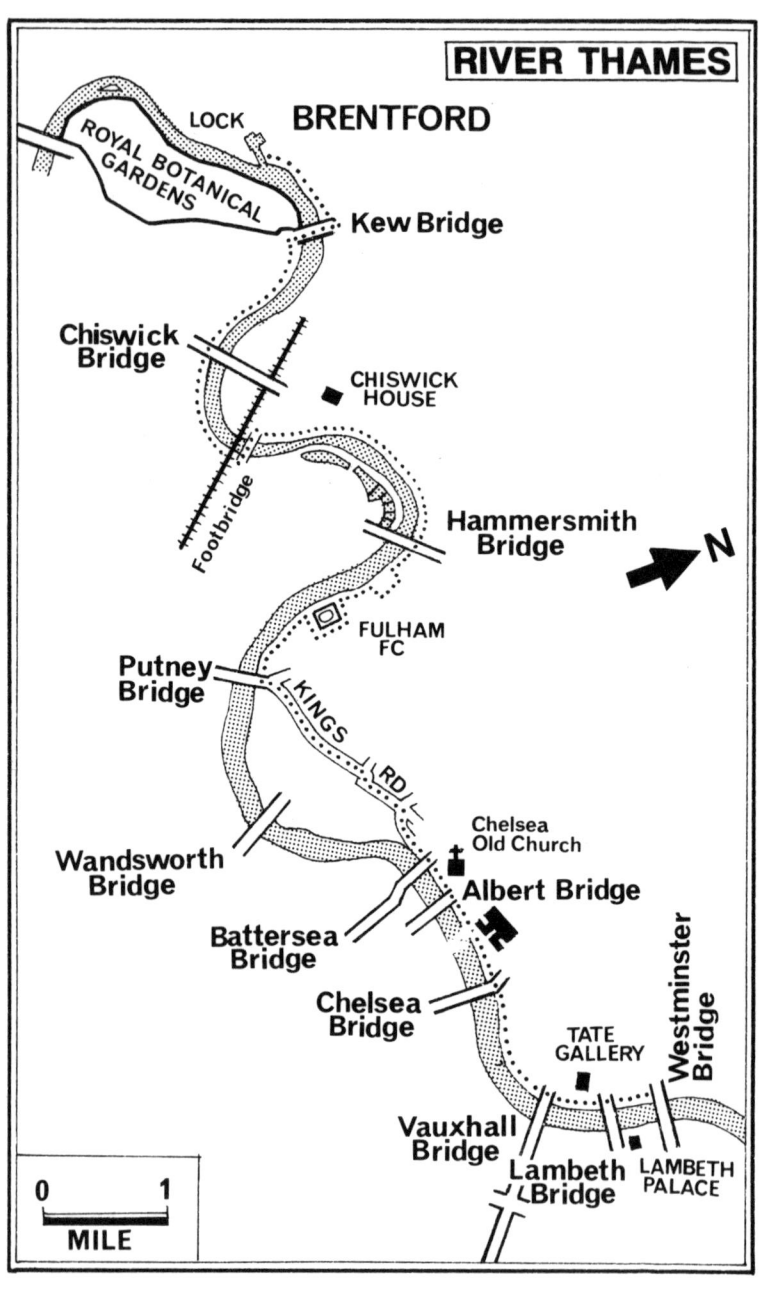

RIVER THAMES

BRENTFORD

LOCK

ROYAL BOTANICAL GARDENS

Kew Bridge

Chiswick Bridge

CHISWICK HOUSE

Footbridge

Hammersmith Bridge

N

FULHAM FC

Putney Bridge

KINGS RD

Chelsea Old Church

Wandsworth Bridge

Albert Bridge

Battersea Bridge

Chelsea Bridge

TATE GALLERY

Westminster Bridge

Vauxhall Bridge

Lambeth Bridge

LAMBETH PALACE

0   1

MILE

1

Westminster to Brentford

12 miles : 19.2 kilometres

Once I had decided to walk from London to York, I knew I just
had to start somewhere by the Thames. Anywhere, it did not
really matter, just as long as it took in a few miles of the country's
greatest river. I could have started by London Bridge, or perhaps
at the Tower, but I quickly settled for Westminster Bridge, within
sight and sound of Big Ben, the great beating heart of England.

The Romans are thought to have first crossed the Thames near
Westminster, and in an attempt to prove that the river had been
fordable at low tide in Roman times, and indeed was still fordable, Lord Noel-Buxton waded in from St Thomas's Steps in 1952.
As in a scene from a Monty Python film, he was soon up to his
neck in the water and had to swim a good deal of the way, no
doubt much to the amusement of the crowds who watched from
Westminster Bridge.

London Bridge was the first of stone to be built across the city's
river. It was begun in 1176 and for nearly 600 years was the only
one, as stiff resistance was sustained against the erection of any
more. Thousands of boats provided the transport up, down and
across the river, and the more bridges there were, the less work
there would be for the Corporation of Watermen. Sense at last
prevailed, and in 1750, after twelve years of work, London's
second bridge, at Westminster, came into use. The opening of the
bridge had been put back three years due to one of the piers
sinking, and after being in use for only a century it was pulled
down and the present bridge, one of the widest and most
handsome in Europe, was opened in 1862, five years after the
Houses of Parliament were completed.

Big Ben first boomed out over London in 1859. It was sounding
the hour as I left the bridge and turned into St Margaret Street in
front of Parliament Square with its craggy, looming statue of

Winston Churchill. Westminster Hall stands on the corner of the square, the only part of the old Palace of Westminster to survive when it was destroyed by fire in 1834. Built in 1097 by William II and rebuilt in 1399 by Richard II, it was then, more or less, the building we see now.

Across the road are St Margaret's, parish church of the Commons, and Westminster Abbey, in which every monarch except the young Edward III has been crowned since 1066.

Several churches have stood on the site of the Abbey, including one at the time of Sebert, King of the East Saxons, in the seventh century. When he died, in 616, he was buried in the church, and later churches, built at the same place, have retained Sebert's tomb. Edward the Confessor built a much better and larger abbey, the first great Norman-style church in England, but he died a week after its consecration in 1065 and was buried in the church he did not live to enjoy. Thus the pattern was set for the abbey, which holds many of the bodies of the kings and queens it has seen crowned.

Unfortunately, I had to leave the venerable temple.

I crossed the road and passed the Houses of Parliament, the statues of Oliver Cromwell and Richard the Lionheart, a magnificent bronze carving of one of England's more romantic kings, astride his horse, his sword held high. Cromwell, too, has drawn his sword, but stands more impassively, a subdued figure, head slightly bowed, the sword pointing into the ground.

Immediately past Westminster Hall I turned into Victoria Tower Gardens, a thin triangle of lawns and trees, overlooked by a statue of Mrs Pankhurst, a woman of the twentieth century. Some might say she was *the* woman of the century for her fight for women's votes. I should think that, throughout her life, the last place she would have expected to be commemorated at was close to the Houses of Parliament.

The gardens end at Lambeth Bridge, opened in 1932 by King George V to replace a bridge built in 1862 which had become suitable only for pedestrian and cycle traffic. There used to be a ferry here from close by the Archbishop of Canterbury's palace across the river. Until the first Westminster Bridge was built, in 1750, it was the only place along the river where horses could be ferried. The ferry was large enough to take a coach and six horses and, when it was discontinued, the Archbishop received £3,000

Westminster Abbey . . . the start of the 304-mile walk

The Houses of Parliament

compensation, because the tolls had been his perquisite. On the
north bank the bridge faces Horseferry Road, named after a ferry
that has not existed for more than two centuries. Lambeth Palace
has been the residence of the Archbishops of Canterbury since
1209, a pot-pourri of periods and styles, with the oldest portion

being the chapel. It was built when the Thames was the great highway of London and it was convenient to have a large riverside residence with its own water-gate. Outstanding features are the Lollards' Tower, built as a water tower in the fifteenth century, and the great gatehouse, with its two impressive towers, one of the few Tudor gateways left in London. Rather more mundane buildings across the water are the Fire Brigade Headquarters and that of the International Maritime Organization.

I walked along Millbank to the Tate Gallery, a pleasing building of 1897 named after Sir Harry Tate, the sugar man who contributed £80,000 towards its cost and also presented £75,000 worth of paintings to go inside it. The Tate stands on part of the site of Millbank Prison, a grim fortress with accommodation for 1,120 prisoners used between 1816 and 1890. A plaque bears the words: 'Millbank Prison 1816–1890. This stone buttress stood at the head of the river steps from which, until 1867, prisoners sentenced to transportation, embarked on their journey to Australia.' Known to prisoners as 'the Tench' – an abbreviation of the word 'penitentiary', presumably, it was a hateful place, and transportation must have been something of a relief after experiencing it. Here now, near where the steps stood, is a sculpted work by Henry Moore, 'Locking Piece', loaned by the Tate Gallery.

Adjacent to the main Tate building is the Clore Gallery, designed by the architect James Stirling and opened to the public in April 1987. It houses a permanent exhibition of works by J. M. W. Turner.

The Albert Embankment, on the other side of the river and running between Westminster and Vauxhall Bridges, was opened in 1869 and named after Queen Victoria's husband, who had died eight years earlier. The famous Vauxhall Gardens were situated near the bridge, behind where the Embankment now stands. They were a place of amusement with 'dark walks and bright lights, music and supper parties'. Vauxhall was highly popular for about 200 years, from King Charles II's time until 1859, opening every evening at five o'clock from about the middle of June to the end of August. Its popularity, until it declined in the middle of the nineteenth century, can be measured by the estimated 16,000 who were there one evening. Tyers Street and Tyers Terrace, which are just off nearby Kennington Lane, are reminders of the man who redesigned the gardens in 1728. River traffic was still extremely important in those days, and vast numbers arrived by boat, landing at a flight of steps leading to the gardens, where admission was a shilling.

Close by, just a short walk down Harleyford Road, is Kennington Oval, the home of Surrey County Cricket Club and one of the most famous Test Match grounds in the world. The first Test ever held in England was staged here in 1880, and the Football Association used to hold its Cup Finals here, long before Wembley Stadium arrived with its capacity of 100,000. Few grounds can boast such a history, yet it does not stand high in most people's lists of favourites. It is a drop of green in an ocean of blocks of flats, a ground of strong character.

The first bridge at Vauxhall was built in 1816 and named Regent Bridge before being changed to Vauxhall Bridge because it led to the pleasure gardens. It was rebuilt in 1906, and each of the four stone piers on which the arches rest is enhanced with a pair of statues, one facing London, one looking away. I crossed the entrance to the bridge and was soon in Pimlico Gardens with their Westminster Boating Base. It was not long before I was taken back to Grosvenor Road, past a restaurant, 'The Elephant on the River', before coming face to face with the huge but not all that unpleasant Battersea Power Station, and its massive chimneys, built in 1933.

The London of the tourist was now behind me. This was working London, even on the river, where the tug *Mersine* pulled a long barge of Greater London Council containers upstream and under the railway line which leads to Victoria. A narrow boat from Pyrford bent on pleasure chugged past, watching carefully as another tug with barges bearing such names as *Audrey* and *Pauline* executed a swift U-turn near the northern bank.

Here is Chelsea Bridge, a suspension bridge and one of London's newest, built only in 1934, and the start of Chelsea Embankment. Here was the site of the Ranelagh Gardens, sister to Vauxhall, a few minutes downstream. Ranelagh was opened to the public in 1742 and later received royal approval with the visit of George III and his family.

Close by, on the Embankment, is a memorial to the officers, non-commissioned officers and men of the VI Dragoon Guards (The Carabiniers) '. . . who gave their lives for their country in the South African War 1899–1902'. A list of those who died of wounds or disease is on one side, another lists those killed at the Relief of Kimberley.

A sign points to the National Army Museum, only 250 yards away and situated in the grounds of Chelsea Royal Hospital, founded by King Charles II and opened in 1682 as a home for old soldiers or those disabled in the wars. Today there are about 500 Chelsea Pensioners. They always look at their proudest in the summer, when they wear brilliant scarlet frockcoats and tricorn hats as against the navy blue greatcoats of the winter, uniforms dating from the early eighteenth century. They are a marvellous sight, the grand old soldiers maintaining a tradition that has been going on for nearly 300 years through the reigns of fifteen monarchs. Meals are eaten in the Hospital's Great Hall, where the Pensioners are served before the officers, a requirement provided by Sir Christopher Wren, who designed the building, to make sure the old soldiers would feel welcomed and honoured. They eat under the watchful eye of their founder, Charles II, who is portrayed on horseback. When the Duke of Wellington died, in 1857, he lay in state in the Great Hall, a fitting resting-place for one of the greatest of England's old soldiers.

The National Army Museum, which opened in 1971, is also housed here, and both it and the Hospital can be visited free of charge. 29 May is Founder's Day; it is also remembered as Oak Apple Day, and the bronze statue of Charles II, which stands in the grounds, is garlanded with branches of oak in memory of his escape after the Battle of Worcester when he hid in an oak tree. Here too is the home of Chelsea Flower Show, held every May.

Across the river stands the attractive Battersea Park, stretching down to Albert Bridge, with its fine boating lake, gardens, magnificent trees and flower beds. Before the original park was opened in the 1850s, the district was known as Battersea Fields, a dark, dismal area that was considered dangerous. The site was marshy and the surface had to be raised by earth from Victoria Dock, which was being built at the time. The Festival of Britain, one hundred years later, brought another change to Battersea. John Piper and Osbert Lancaster designed the new Festival Gardens for the Festival of Britain, a modern version of the pleasure gardens which existed at Vauxhall. Even its popularity declined with time, of course, and the amusement section was closed in 1975, leaving a site now used for various events from ballet to exhibitions. In 1985 the nuns and monks of the Japanese Buddhist Order Nipponzan Myohoji built the magnificent Peace Pagoda which now stands in the park very close to the river. With its 33-foot pinnacle decorated with bronze and gold it makes a wonderful exotic landmark across the water.

Albert Bridge, with its lovely lattice work giving it a cobwebbed appearance, was now in sight, but before it came the house where the Marquess of Ripon (1827–1909), a Viceroy of India, once lived. Close to the bridge is Cadogan Pier, with its boats bobbing about in the water which was fast running back to the sea when I was there. If that was not soothing enough, standing by the Thames was a rather fine carving of a nice young lady, her hair swept back by the wind, an example of the work of George Derwent Wood RA.

Here is Cheyne Walk, home of artists, poets and writers since it was built early in the eighteenth century, a delightful stretch by the river. Turner, the celebrated painter, lived here, so did the railway-building Brunels, while Thomas More, Lord Chancellor to Henry VIII, had a mansion in the neighbourhood with grounds running down to the Thames. A pleasant statue of Sir Thomas More (1478–1535) shows him sitting piously, almost beseeching-ly, his hands clasped and knees together, in front of Chelsea Old Church. The church was built early in the fourteenth century (More, who attended the church regularly, built a chapel here for his first wife), partly demolished in 1667, partly restored in 1910 and badly damaged during an air raid in 1941. Much of More's chapel survived but restoration was necessary for what is now a beautiful church, still crowded with history.

A man tootled away on a saxophone as I closed in on Battersea Bridge, timber-built in 1772 and a worthwhile subject for several famous painters before being replaced in 1890 by the present one. Nearby was another of those famous houses, this time where Sylvia Pankhurst, 'Campaigner for Women's Rights' (1882–1960), lived.

I was soon at Chelsea Wharf, with its huddle of houseboats, and into Lots Road, where I came across the pub with arguably the longest name of any in the country. At first sight I thought it was simply named 'Up the Creek', then 'In the Balloon up the Creek', until I turned the corner to find the full and fascinating

Statue of Sir Thomas More in front of Chelsea Old Church
alongside the Thames

title, 'The Ferret and Firkin in the Balloon up the Creek'. There.
Beat that if you can. There are plenty of other Firkin pubs in
London, all selling strange-sounding but big-hearted beers that
warm the cockles of your heart.

I soaked up the flavour of Kings Road, only a few minutes' walk

away from the river, and revelled in the beautiful sight of Con-
corde, her nose sniffing out Heathrow, as I made my way to
Putney Bridge and back to the river. It was here that the first
bridge upstream of London was constructed, a wooden toll-
bridge erected in 1729, rebuilt in 1886 and widened in 1933.

I approached the river by way of a sign that points to Willow
Bank and River Walk, the start of a beautiful stretch of river-
walking. This is Bishop's Park, the name of the grounds belonging to Fulham Palace, the summer residence of the bishops of
London for nearly 900 years until they moved out in the 1970s.
The well-kept path by the river is lovely, tree-lined and lit by
lamps with round shades like footballs, perhaps in deference to
the approaching Fulham FC ground. A 'fitness trail' stands in the
parkland alongside jumpingjacks, hopkicks, touchtoes and
kneelifts; a small lake is adorned with weeping willows, and the
river is satisfyingly close. Two single oarsmen pulled upstream,
overtaken by half a dozen fast-moving flying ducks; *Princess Freda*
chugged down towards London, and a double-decker boat glided
by.

The path took me round Fulham's football ground, the nearest
League ground to the Thames and which looks, in parts, as grim
as any penitentiary. It could easily have been a prison, heavily
guarded with floodlights, and it was nice to return to the river,
where about twenty ducks were all foraging for food in the mud
at the water's edge, heads down like an examination class.

Hammersmith Bridge and the beflagged twin towers of Har-
rods Furniture Depository were in sight as I passed attractive,
new property before another diversion took me away from the
river and round Crabtree Tavern, a pleasant-enough pub, I am
told, but one to avoid any Saturday lunchtime when Fulham are
playing at home. The way to the bridge is past Thames Wharf
Studios before following a cycle sign for 'Hammersmith ½' which
leads back to the river.

Hammersmith suspension bridge, opened in 1887, is one of
London's most handsome bridges, attractively garnished but
rather narrow for today's traffic. Still, it is nice to look at, and I
lingered a while before approaching another lovely stretch alongside the river. I walked past the Blue Anchor and Rutland pubs,
some two-bedroomed luxury apartments and the Furnivall Sculling Club before reaching Dove Marina where a plaque informed

Hammersmith Bridge, one of London's most handsome bridges

me: 'Here was the Creek, a harbour where the village of Hammersmith began: until 1929 sailing barges entered the harbour to discharge their cargoes and 200ft to the north was the High Bridge across it. This garden overlooks Hammersmith Reach, long famous for sailing and rowing. It is named after Dr Frederick James Furnivall (1825–1910), man of letters, social reformer, and founder of the Furnivall Sculling Club.' This is a popular boating area, with Hammersmith Boathouse, Sons of the Thames Rowing Club and London Corinthian Sailing Club leading to the 1978 Pub of the Year, the Old Ship, a lovely pub in a lovely spot.

Hammersmith Terrace is full of splendid terraced houses and cottages, and those familiar blue plaques tell us of Edward Johnston (1872–1944), master calligrapher, who lived here 1905–12, and Sir Emery Walker (1851–1933), typographer and antiquary, who lived here 1903–33. A. P. Herbert, the author, who set his novel *The Water Gypsies* by the river near Hammersmith, and who died in 1971, lived here for more than fifty years, at numbers 12 and 13.

The river turns into a creek near the road as a small island, Chiswick Eyot, separates the water from the river. Soon it is back to the street, down Pumping Station Road, before picking a way through the blocks of flats to get back to the river. The planes for Heathrow were non-stop, as busy as the swallows who were jostling for air space as they gorged themselves on midges. Joggers brushed through the willow trees on a terraced stretch, near where stands Chiswick House, built in Italian style in 1725 by the Earl of Burlington as an artistic showplace but later converted into a home.

For the first time since leaving Westminster, I crossed the river, using the footbridge that runs alongside the railway bridge at Barnes. I walked in front of the White Hart Hotel, a lovely old Victorian pub, and the smaller Ship Inn, on my way to Chiswick Bridge, near where the University Boat Race, that curious piece of English tradition, ends after starting at Putney. There seemed no way of telling which side of the river would provide the better walk but, having only just crossed to the south side, I decided to stick with it. I was rewarded with a lovely, peaceful walk through the trees, although I was left gazing across the river at the picturesque Strand-on-the-Green, a quaint waterside hamlet, once a fishing village and now an appealing spot. The City Barge is a popular pub, built nearly 500 years ago and originally called the Navigator's Arms before taking its name from the City of London boat that used to inspect barges and collect dues.

It took me just under half an hour to walk from Chiswick Bridge to Kew Bridge, where I crossed back over to the right bank so I could approach Brentford and the start of the Grand Union Canal. I rejoined the river along The Hollows and was then forced up onto Brentford High Street.

Kew Gardens, which became a botanic garden late in the seventeenth century, and Kew Palace, built in 1631 by a Dutch merchant, stand on the opposite side of the river. On the Brentford side is Syon Park with its House, a convent built in the fifteenth century, seized by Henry VIII and used to imprison Catherine Howard before her execution. James VI and I gave Syon to the Earl of Northumberland and today, after nearly 400 years, the property is still in the family.

Ferry Lane and Dock Road lead back to the Thames and its junction with the Grand Union Canal. Large tidal locks join them, two of the greatest waterways in the country. I turned my back on the Thames, the magnificent Thames, and Kew Gardens, directly across the river, and set my face to the north.

OS MAP 176 or London A–Z
Information Centres: Tel. 01-730 3488 or 01-940 9125

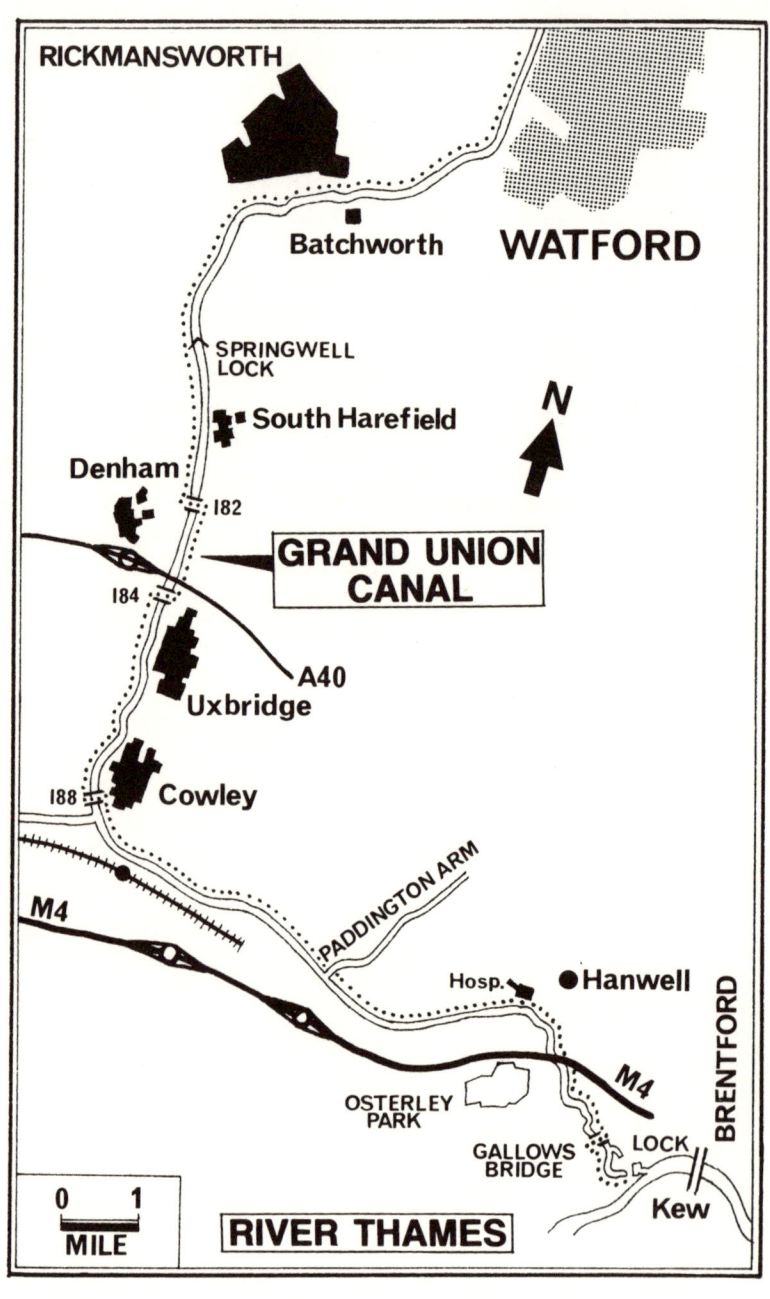

2

Brentford to Watford

21 miles : 33.6 kilometres

The Grand Union Canal is composed of several separate canals, each of which used to be independently controlled. It was formed in 1929 by the union of ten canals, the most important of which was the Grand Junction Canal which had been created to form a more direct link between the Midlands and London. The Grand Junction, in fact, forms the next eighty-nine miles of this walk, from Brentford to Norton Junction, where it forks left to Braunston but where I forked right for Leicester and Nottingham.

The Oxford Canal, seventy-seven miles long and completed in 1790, joined with the River Thames to form the main waterways route between the Midlands and the south for a short time in the closing years of the eighteenth century. It was much too long and tedious a journey, and it was obvious that a more direct route was needed for transporting goods. A survey was quickly carried out and the present route was authorized in 1793, despite several physical obstacles. There were to be two summit levels, 101 locks, two tunnels and quite a few aqueducts.

Blisworth Tunnel held up completion for several years, and a tramway had to be built over the hill to connect the two parts of the canal until the tunnel was finished. It was an enormous job in the days when there was no mechanical help, just men and horses and sweat. Completion arrived in 1805, twelve years after the start, but a new problem appeared a year later, when the embankment leading to the stone aqueduct over the Great Ouse at Wolverton burst. The damage was repaired but three years later the aqueduct itself collapsed and the present iron trough that straddles the valley was erected in 1811.

Many branches were built off the main line, running away to Northampton, Buckingham, Aylesbury, Slough, Wendover and Paddington. In 1894 the Grand Junction absorbed the Leicester-

shire and Northamptonshire Union and the original Grand
Union, which had two long tunnels and a flight of ten locks in its
twenty-four miles. The others were gathered in in 1929, the
Regent's in London, the Warwick and Birmingham, Warwick and
Napton, Birmingham and Warwick Junction, Leicester, Loughborough and Erewash Canals and Navigations. All now are one,
more than 300 miles of canal, and all are navigable. The oldest
among them was the tiny Loughborough Navigation, just over
nine miles long, mainly canalized river and opened in 1778.

Brentford is a place of history, yet there is little to stir the senses
in a town known more for its industry. Once it was important, the
capital of Middlesex. Iron Age Britons lived in the area, Julius
Caesar is thought to have crossed the river here, and King Canute
fought here. Another battle took place during the Civil War,
when the town was severely looted by the Cavaliers. George I
had a fond affection for the place because it reminded him of his
native Hanover. Two of the four Anglican churches in the town
are no longer used for worship and outside one of them, St
Lawrence's, is a reminder of the Grand Junction days, with an old
drinking fountain given by the Grand Junction Water Company
bearing the words: 'Let him that is athirst come and whosoever
will let him take the water of life freely.' Another church, St
George's, houses the Piano Museum, which contains an interesting collection of mechanical and unusual instruments, automatic
pianos and violins and nickelodeons. The Grand Junction pops
up again in the High Street, where the old pumping station now
contains a living steam museum.

The Butts, a tree-lined square with eighteenth-century houses,
is worth seeing. On Butts Common, in 1558, six martyrs were
burned at the stake for their religious convictions. The entrance to
Syon House is two minutes walk from the bridge that takes the
High Street traffic over the canal. It is set in open parkland and for
the dedicated walker there is a public right of way across the park,
following the last journey of Henry VIII as his body was transported from London to Windsor for burial.

The next lock after the one linking with the Thames comes
immediately after the High Street bridge. It is called Brentford
Gauging Locks, and this was where traffic joining the canal from
the Thames used to be assessed for tolls. It was a major trans-
shipment centre for goods from the north when the first section of

the canal was opened between Brentford and Uxbridge in 1794.

The towpath passes under the great expanse of a covered dock, quite spooky in a strange way, as it leads under the Great West Road to the M4 motorway carrying traffic thundering its way between London and the West Country. The motorway runs alongside the canal for a while from Clitheroe's Lock, directly opposite which is Boston Manor, a distinguished house, red-brick and pleasing to the eye, which was built in 1623. This is a pleasantly wooded area, leading up to the old iron bridge, intriguingly named Gallows, which carries the towpath to the other side. The bridge is believed to have been the first canal bridge built by Horseley Iron Works, near Birmingham, and carries its name with the date 1820.

The path goes under the Piccadilly underground line, under the motorway and through an area thick with elderberry. Planes bound for Heathrow, five miles away, share the air space with swarms of swallows, as the path goes by footbridge over the River Brent, which has been canalized from Brentford. Osterley Lock is a reminder that yet another impressive building, Osterley House, is just around the corner. Well, twenty minutes walk away, to be more exact.

Osterley House was built by Sir Thomas Gresham in the middle of the sixteenth century and had the great distinction of enter-taining Queen Elizabeth. The Queen, probably making light conversation over dinner, suggested that perhaps the central courtyard might appear more handsome with a wall across it. Her suggestion was taken as a command and by the morning it had been obeyed: masons had been up all night building it. Robert Adam took part in the reconstruction, which took several years in the eighteenth century, and much of the furniture he designed for the house is still there. There are lakes and fields and a walk through the grounds of a beautiful property which is now in the hands of the National Trust.

Back on the canal, it was around here that I came across an overgrown wharf with a gun ranged on the canal. It was quite big, with a barrel nine or ten feet long and still capable of swivelling to aim up or down the canal. Nobody I spoke to knew what it was, but if it was a relic of the Second World War it was certainly wearing well. I moved on up to the Hanwell flight of locks, six of them, standing alongside St Bernard's Hospital, which used to be an asylum and which is remembered in the name of one of the locks – Asylum Lock. Close by is the arch, now bricked up, that formed the entrance to the hospital's own dock. Built into the towpath near the middle lock are the horse-steps which allowed any animal dragged into the water by the tight bend to climb out again. Joggers panted by, as joggers do, and a gathering of goats across the water watched with an exaggerated air of disbelief and boredom.

Hanwell is just over to the right, and close to the railway station is the handsome viaduct, built in 1838 and designed by Brunel, which carries the railway over the River Brent. The views are lovely, and it is said that any train carrying Queen Victoria would travel slowly across the viaduct so that Her Majesty could savour them. My view was not quite so impressive, although just beyond Three Bridges, where canal, road and railway come tangling together, is a pleasant stretch around Norwood Locks, river-like with trees hanging over the water and boats tucked beneath them. The number of miles to the Thames is displayed on a stone post, and just beyond the pub, the old Oak Tree, is a row of houses called 'Industrious Cottages'.

Bull's Bridge junction signs the turn-off to Paddington, thirteen miles away, and a cottage garden contains many canal souvenirs, including a notice from the Warwick and Napton Canal Navigation warning traction-engine drivers and others that the bridge was insufficient to carry weights beyond the ordinary traffic of the district. For 200 years the Bull's Bridge site has been an important maintenance centre, and it is well worth a few minutes lingering and looking. Then straight on for the north through a few tantalizing smells as I left the Nestlé factory on the left and Wall's meatworks on the right. What a combination.

Soon after West Drayton railway station comes the Slough Arm, five miles of one of the last canals to be built in Britain. It was built in 1882 to serve the many brickfields and gravelpits in the vicinity, some of which turned into tips for refuse around the turn of the century. Nearby, at Cowley Peachey – the name was derived from the Peche family who were local landowners in the thirteenth and fourteenth centuries – is the Paddington Packet Boat Inn, which recalls the twenty-five-mile Paddington to Cowley passenger service, opened in 1801 and closed after only five years.

I was happy to reach Uxbridge. I had been there a few weeks earlier for a Middlesex county cricket match, and I enjoyed the lovely old place, an ancient market town that became an important coaching centre on the road between London and Oxford. My return meant I could again visit the Crown and Treaty Inn built in the first half of the sixteenth century when it was known as Place House. It was much larger in those days, and the part remaining is only one third the size of the original.

In 1645, during the Civil War, Parliament and the Crown tried to come to terms, on the initiative of King Charles I after his defeat at Marston Moor. The talks were held at Uxbridge and were conducted by thirty-two Commissioners, sixteen from each side. The King's Commissioners took the Crown Inn on the south side of the High Street for their headquarters, and the Parliament men the George Inn on the other side of the road. A room in the existing building is, by tradition, believed to be the one in which the treaty was discussed – and then abandoned – but Carolynne Hearmon, who wrote the book *Uxbridge* in 1982, thinks it unlikely and that the middle room was probably demolished in the eighteenth century. The room remaining, she feels, is more probably one the Commissioners retired to.

Two and a half years after the treaty discussion failed, the Parliament Army made its headquarters in Uxbridge, and Oliver Cromwell, perhaps in a frenzy of one-upmanship, chose to stay at the Crown Inn, the Royalists' headquarters for the talks.

During the eighteenth century the main block and one wing were demolished and the remaining building was divided into several tenements. Improvements to the High Street in 1785 meant that the line of the road was taken straight through the Treaty House grounds, and soon after that the building became an inn with stabling for over forty horses.

Another dramatic event occurred at Uxbridge in August 1555, when three Protestants were burned at the stake for their faith and as an example to the Uxbridge people. In Foxe's *Book of Martyrs* is a graphic description of the death of one of them, John Denley, a gentleman from Kent who had been tortured by the Bishop of London but refused to deny his faith. 'Being set in the fire with the burning flame about him, he sung in it a psalm. Then cruel Doctor Story, being there present, commanded one of the tormenters to hurl a faggot at him, whereupon being hurt there-

The Fisheries Inn on the Grand Union Canal at Copper Mill
Lock near Harefield

with upon the face that he bled again, he left his singing and clapt both his hands on his face.'

John Leland, Henry VIII's antiquary, had visited Uxbridge in the 1540s and reported: 'The whole town lieth from the West rising a little to South-East. In it is but one long street; but that for timber is well builded. There is a celebrated market once a week and a great fair once a year at the feast of St Michael. There be two wood bridges at the west end of Uxbridge town and under the westernest goeth the main arm of the Colne river. The lesser arm of the Colne goeth under the other and each of them serves there a great mill.'

There was no more important event in Uxbridge's history than the opening of the Grand Junction Canal. The section from the Thames to Uxbridge was finished in 1794 and opened on 3 November with a grand procession from Brentford. The *Northampton Mercury* wrote: 'The opening of this part of the canal was celebrated by a variety of mercantile persons of Brentford, Uxbridge, Rickmansworth and other vicinities, forming a large party, attended by a band of music, with flags and streamers, and several pieces of cannon, in a pleasure boat belonging to the Corporation of the City of London, preceding several barges laden with timber, coals, and other merchandise to Uxbridge.'

Soon after leaving Uxbridge, the canal passes Denham, famous for its film studios, and crosses over the meandering River Misbourne. The Colne Valley passenger boat was standing idly on the other side of the canal, being tidied. I asked where it ran to. 'Down to Uxbridge and Brentford, up to Fishery,' I was told. I took 'Fishery' to be Fishery Lock, sixteen miles away. A ride would have been nice, but this was autumn and there were no more trips, I was told, until April.

Several lakes, formed by sand and gravel workings, line the canal for the next five or six miles. As I crossed the canal on a roving bridge (designed to enable horses to cross the canal without removing the tow rope) I glimpsed the reds, blues and whites of the boat sails and heard a few excited screams. Close to Bridge 180 near South Harefield is a lagoon, now operated by Harefield Cruising Club, where twenty-four working boats were sunk in 1958, when commercial carrying was nearing its end.

The Horse and Barge pub on the canalside carries an engaging swinging sign, leading to Widewater Lock, near where several new houses stand close to the water, and a wartime defence pillbox incongruously looks over the canal from the bottom of somebody's garden. Before reaching Black Jack's Lock another huge lake, Harefield Flashes, appears on the left, a lovely scene with boats and country, a few local planes drifting about and delightful views opening out. Black Jack's Lock takes its name from a Negro employed by a local landowner to harass boatmen passing through at night. When Black Jack was murdered, a haul of windlasses was found in the hollow of a nearby tree. Now he takes revenge by haunting the place; even so, it is still a lovely spot, with a restaurant housed in the old cornmill and a thatched house nearby.

As I passed Fishery Cottage with its sweet peas climbing a wire fence, I heard the sounds of a football match drifting down from the hill, perhaps from Harefield. The nearby Fisheries Inn has an interesting sign showing a young man who caught, and kept, a whopper of a fish. I wondered . . . was this the 'Fishery' the Colne Valley trip boat came to? Five minutes after passing Copper Mill Lock, where Waterways Cottage stands between the canal and the River Colne and where canoeists play in the weir, I saw an old Grand Junction Canal signpost 'G.J.C.Co. Braunston 77 miles'.

I pressed on, to Springwell Lock, with Harefield Pottery, boasting hand-made stoneware pottery and welcoming visitors, nearby. Close by was a boat, *Fulbourne*, built for the Grand Union Canal Carrying Company in 1937 and known as a 'large Woolwich'. It is believed to have worked into the 1960s before being converted into a pleasure boat.

The area around Stocker's Lock is a picture, absolutely beautiful, with its lock cottage, Georgian house and nestling farm. Several boats were understandably moored close by. It was late afternoon but still warm enough for a cat to snooze peacefully alongside a yellow bird in a cage on the roof of a boat.

Moor Park golf course stands across the canal close to Batchworth and is notable for its magnificent club house built in the early eighteenth century. Cardinal Wolsey was one of the owners of Moor Park before it came to the Duke of Monmouth, son of Charles II, who fought a terrible losing battle at Sedgemoor in 1686. He started building the house that exists today, standing at the head of the park, a splendid setting. Across the water, not far from Lot Mead Lock, is the Tithe Barn at Croxley Hall Farm, the second largest in the country. It is 101 feet long and 38½ feet wide, with each side divided into ten bays. It used to be nick-named 'Wolsey's Slaughterhouse', a reminder of the feastings when Cardinal Wolsey lived nearby at the Palace of the Moor. The roof caved in and the building was in a poor way in the 1960s but has since been restored.

Rickmansworth marks the meeting of the waters, the rivers Colne, Chess and Gade, giving this pleasant town an added quality. There is a lot to commend the stretch past Croxley Green towards Watford: the beauty of the countryside, the accompanying river, the perfect quiet. A squirrel hung upside down in a hedge, munching away to its heart's content and unaware of the intruder on the towpath. Quite, quite beautiful.

Watford quickly followed, the biggest town in Hertfordshire, a busy old place with a large, modern shopping area and the old part of the town grouped round the church, which has parts dating from the thirteenth century.

This had been an interesting, at times extremely beautiful, stretch of canal with many lakes, rivers running alongside and a good, clean, dry path to see it all from.

OS MAP 176 or London A–Z (Part)
Information Centres: London (01-730 3488 or 01-940 9125); Uxbridge (0895 50706); Rickmansworth (0923 776611)

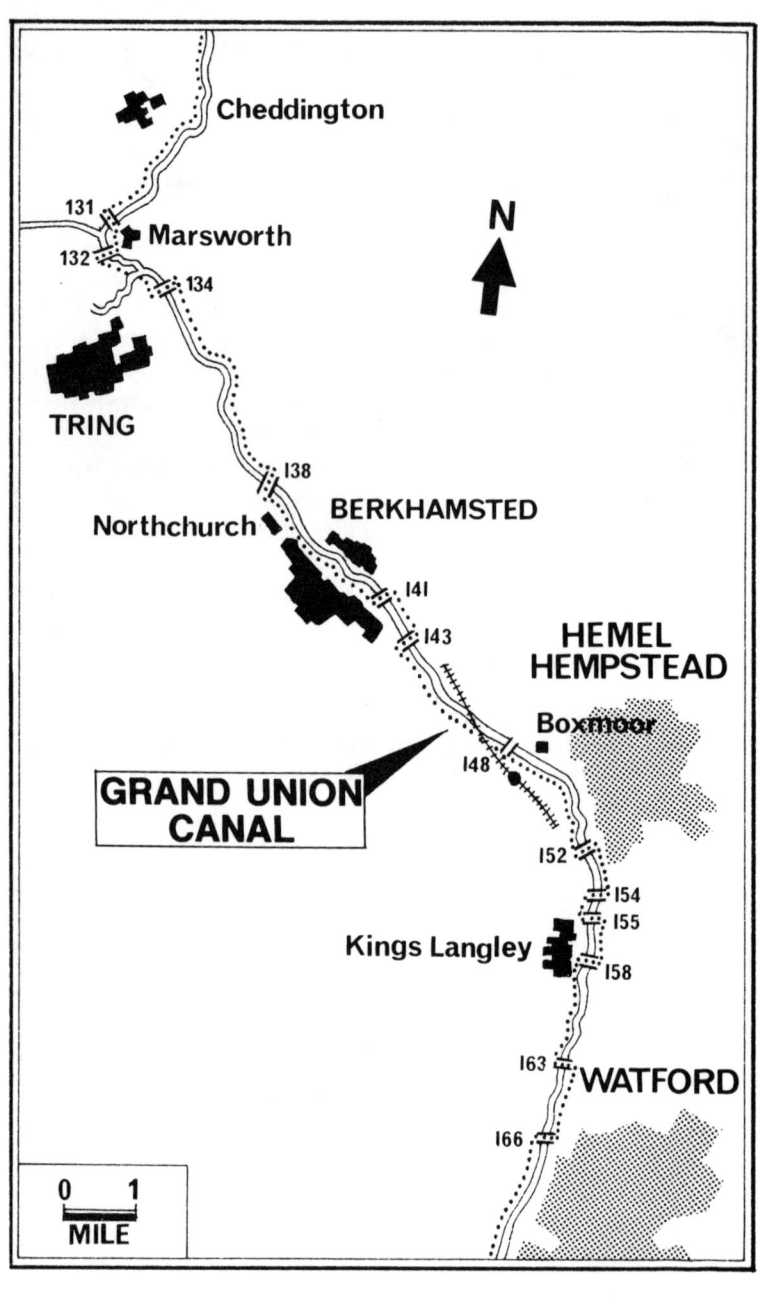

3

Watford to Cheddington

20 miles : 32 kilometres

Watford is well blessed with parkland and playing fields, hundreds of acres of them. One of the loveliest pieces of open land is at Cassiobury Park, which gives about an hour's delightful walking by the Grand Union Canal. The Capel family of Cassiobury were rewarded with the earldom of Essex in return for their efforts in the Civil War, and this was once their land – but no longer: it is now the haunt of walkers, picnickers, boaters and golfers.

It was quite early in the morning when I set off, misty but promising. A train on the Metropolitan Line rolled high overhead towards Watford, one of the outposts of the London tube system and the last bit of town noise for about three miles. Squirrels, wood pigeons and anglers shared the lovely surroundings leading to Ironbridge Lock and Cassiobury Park Lock, with its whitewashed cottage, decorated with geraniums in pots on the windowsills and hanging baskets of flowers. Fine old trees line the route, affording pleasant walking all the way to Lady Capel's Lock before the parkland runs out near the trunk road, the A41.

Grove Mill, on the other side of the waterway, features old agricultural machinery on its lawn, and just before the arrival of the traffic there is a signpost – 'Braunston 70 miles'. Near Hunton Bridge, where the A41 crosses, is a stone to commemorate E. G. Christopher, who died on 1 January 1970, and Charles Curran, who was killed on 7 June 1970, during the construction of the Gade Valley Trunk Sewer.

The land around here belonged to the earls of Clarendon, neighbours and relatives of the Capels. Grove Park was theirs, and it was in this vicinity that the towpath, which has been on the left-hand bank for ten miles, is forced to the other side, part of the Earl of Clarendon's conditions for allowing the canal through his property. A sign across the canal points to the King's Head at Hunton Bridge, a well-positioned pub directly opposite the gates of Benskin Brewery. The Dog and Partridge is not far away, in Old Mill Road, and here too is the King's Lodge restaurant, dated 1642, a low building, extended in the nineteenth century, which was a hunting lodge for Charles II. This is a peaceful canalside village on the road from Kings Langley to Watford, a pleasant dropping-off place.

Back on the towpath, large willow trees dipped their heads in the water, and on the pathside was a huge field of corn on the cob. A railway line runs away on the right, presumably bound for London, and ahead is the M25 motorway, newly constructed, gracefully fanning across the valley and looking as clean and well scrubbed as a boy on his first morning at school. Home Park Mill Lock leads into Kings Langley, and just beyond it on the towpath side of the canal was the unlikely sight of an aircraft looking out across a lake.

'It was a Sea Vixen which served in the Falklands,' a man told me. 'It cost £10,000 and now, three years later, it's bringing offers of £20,000. A local contractor made the lake. He'd always wanted a plane, and there it is. The wings were taken off, and it was transported here from Farnborough. A chap in Watford has three, you know.'

I suggested he might next get an aircraft-carrier for the lake. 'I don't think he'll go that far,' said the man. He thought for a moment, then added thoughtfully: 'Though he does get funny notions.'

Kings Langley was coming up, the home of kings in medieval times, though only a fleck of the palace is left today. The body of Richard II was brought here after his death but was later removed and taken to Westminster, my starting-point nearly forty miles earlier. The canal here follows the course of the River Gade, wide and deep, a country river, and passes the Ovaltine factory on its way to King's Langley Lock and Nash Mills Lock. The spire of St Mary's Church, Apsley, reaches up to the sky, and the A41 to London, Aylesbury and Hemel Hempstead comes up close.

Apsley Bottom Lock heralds the arrival of Hemel Hempstead and Boxmoor on the right. A boat called *Nile* was going through the lock with a Scottish terrier on board. A perspiring woman leaned on the arm of the lock, straining to get it open. Her husband leaned on the rudder of the boat. 'Push up, Doris,' he called. The River Gade, which had mills alongside it in Saxon times and delicious eels in it in the days of the Domesday Book, goes off to the left before reaching the middle of the three locks. Just beyond the top one, by Bridge 152, is a good-looking Benskins pub, the Albion, invitingly situated alongside the canal with garden seats by the water.

About half a mile away from the next bridge is Hemel Hempstead, which has a church of striking Norman construction with a beautiful 200-feet-high spire, which was added on a little later. The church is rich in treasured pieces of history such as an Elizabethan chalice, a sixteenth-century chest, a medieval stone coffin and an edition of Foxe's *Book of Martyrs* dating from 1610. A couple of grand old pubs are the Sun Inn, dated 1726, and the King's Arms, more than 300 years old.

The river crosses beneath the canal twice around Boxmoor, and I watched, as ever dazzled by the bird's blaze of colour, as a kingfisher flew out and up the canal. It headed towards Fishery Inn, built in the reign of William IV, ideally situated with tables by the water and with huge signs proclaiming '30 miles from Brentford' and '64 miles from Braunston'.

It was just beyond here that I experienced one of those chance happenings that make any day memorable.

Two boys were playing on their bikes by Bridge 148, the one after Fishery Inn. One wanted to go for a spin along the towpath; the other grabbed my attention by talking of going to a nearby grave. I asked what it was and was told it was a highwayman's – would I like to see it?

I followed the boys off the bridge and under the railway line before turning right through a gate into a field. Now they were not sure which way to head. One thought they should make for the Swan Inn, a few hundred yards away to the left, on the other side of the A41. The other wanted to go right. He was looking for a group of chestnut trees. There were clusters of chestnut trees about, which confused him, but he felt they ought to go right and nearer the road. I told them to shoot off on their bikes and if they found it to give me a call. A few minutes later they were waving and I made my way to a second field and the chestnut trees a hundred yards from the road. Just beyond the trees was the small headstone and footstone of 'Robert Snooks 11 March 1802'.

Watford Ironbridge Lock

The boys, Julian Wheals and Andy Double, both twelve-year-olds, told me that Snooks had been hanged in one of the trees and that the marks of the rope could be seen. They told me, too, of the legend of Robert Snooks: 'If you come here at midnight and walk round the grave twelve times, you will see him hanging from the tree. Some have seen him standing by the grave and he has been seen on his horse near the road.' I wondered how much was fact, how much was boys' fantasies, and as soon as I reached home I went to Manchester Central Library to discover all I could about Robert Snooks.

I started with Arthur L. Hayward's *Complete History, Lives, and Robberies of the Most Notorious Highwaymen* and tried Patrick Pringle's *Stand and Deliver* before turning to *Half Hours with the Highwaymen* by Charles Harper, written in 1908. There I found the story.

Robert Snooks, a native of Hungerford, was living in Hemel Hempstead, in the immediate vicinity of Boxmoor, in 1800.

He had often observed the post boy carrying the well-filled mail bags across the lonely flat of Boxmoor and (he is described as having been of remarkably fine proportions) thought how easy a thing it would be to frighten him into giving them up [wrote Mr Harper]. Accordingly, on one sufficiently dark night, he waited upon the moor for the post boy, stopped him and, adopting a threatening demeanour, instructed him to carry the bags to a solitary spot and then go about his business. The frightened official immediately hurried off to the post master of the district, one Mister Page of the Kings Arms, Berkhamstead, and told his tale, leaving Snooks to ransack the bags and take what he thought valuable. The bags, turned inside out, were found the next morning with a heap of letters, torn open and fluttering in all directions across the fields. It subsequently appeared that the highwayman had secured a very considerable booty, one letter alone having contained £5 in notes. The post boy did not know the man who had terrorised him: only that he was a 'big man'; but the simultaneous disappearance of Snooks left no reasonable doubt as to who it was.

This was Snooks's first essay in the dangerous art and it proved also his last. Hurrying to London, he took up his abode in Southwark and presently had the dubious distinction of reading the reward bills issued, offering £300 for his capture. After a while he thought himself comparatively safe and was emboldened to make an effort at negotiating one of the notes he still held. Afraid to do this in person, he thought he might see what would happen if he tried to pass one of the notes through the intermediary of the servant of the house where he was lodging and accordingly sent her to purchase a piece of cloth for a coat, handing her a £5 note. The tradesman evidently found something suspicious about the note thus tendered and returned it, with the message that 'there must be some mistake.' Whether the tradesman would have followed this up by communicating any suspicion he may have had to the authorities does not appear.

Snooks fled to Hungerford, his birthplace, and evaded capture until 1802, when he was arrested, on the information of a post boy who had been at school with him.

He was in due course brought to trial at Hertford Assize, found guilty and sentenced to death. It was judged expedient, as a warning to others, that he should be executed on the scene of his crime, the selection of the spot falling to Mister Page who, besides being postmaster of Berkhamstead, was High Constable of the Hundred of Dacorum. As a further warning, and one likely to be of some permanence, it was originally proposed to gibbet the body of the defunct Snooks on the same spot so that, swinging there in chains on the moor, it might hint to others the folly of doing likewise. But the time was growing full late for such exhibitions; the inhabitants of the district protested and this further project was abandoned. Journeying from Hertford gaol on the morning of the fatal March 11, 1802, Snooks, according to a surviving tradition, was given a final glass of ale at the Swan Inn, at the corner of Box Lane, and is said to have remarked to the rustics hastening to the scene of the execution: 'Don't hurry; there'll be no fun till I get there.' The usual large and unruly crowd, that could always be reckoned upon on such melancholy occasions, was present, and seemed to regard the event as no more serious than a fair. To those thus assembled, Robert Snooks, standing in the cart under the gallows, held forth in a moral address:

'Good people, I beg your particular attention to my fate. I hope this lesson will be of more service to you than the gratification and the curiosity which brought you here. I beg to caution you against evil doing, and most earnestly entreat you to avoid two evils, namely "Disobedience to parents" – to you youths I do particularly give this caution – and "The breaking of the Sabbath." These misdeeds lead to the worst of crimes: robbery, plunder, bad women, and every evil course. It may by some be thought a happy state to be in possession of fine clothes and plenty of money, but I assure you no one can be happy with ill-gotten treasure. I have often been riding on my horse and passed a cottager's door whom I have seen dressing his greens and perhaps had hardly a morsel to eat with them. He has very likely envied me in my station who, though at that time in possession of abundance, was miserable and unhappy. I envied him and with most reason, for his happiness and contentment. I can assure you there is no happiness but in doing good. I justly suffer for my offences and hope it will be a warning to others. I die in peace with God and all the world.'

The horse was then whipped up, the cart drawn away from beneath the gallows tree and Robert Snooks had presently paid the harsh penalty of his crime. He had behaved with remarkable courage and espying an acquaintance in the crowd, offered him his watch if he would promise to see that his body received Christian burial. But the man, unwilling to be recognised as a friend of the criminal, made no response, and Snooks's body was buried at the foot of the gallows. A hole was dug there and a truss of straw divided. Half was flung in first; the body upon that, and the second half on top. The hangman had half stripped the body, declaring the clothes to be his perquisite, and would have entirely stripped it, had not the High Constable interfered, insisting that some regard should be had to decency. A slow-moving feeling of compassion for the unhappy wretch took possession of some of the people of Hemel Hempstead who, on the following day procured a coffin, reopened the grave and, placing the body in the coffin, thus gave it some semblance of civilised interment; but, those being the times of the body snatchers, doubts have been expressed of the body being really there. It is thought that the body snatchers may afterwards have visited the lonely spot and again resurrected it. Two rough pieces of the local plum pudding stone were afterwards placed on the grave and remained until recent years.

Boxmoor is not now the lonely place it was. The traveller who seeks Snooks's grave may find it by continuing northward from Apsley End, passing under the railway bridges and coming to a little roadside inn called The Friend at Hand. Opposite this, on the right hand side of the road and between this road and the railway embankment, runs a narrow strip of what looks like meadow land, enclosed by an iron fence. This is really a portion of Boxmoor. At a point 150 yards past the inn look out sharply for a clump of five young horse chestnut trees growing on the moor. Close by them is a barren space of reddish earth with a grassy mound, a piece of conglomerate, or 'pudding stone', and a newer stone inscribed with 'Robert Snooks, 11 March 1802'. This has been added since 1905 and duly keeps the spot in mind.

So it was not the fantasy of two young boys.

I returned to the canal at Winkwell, in time to drop in at the Three Horseshoes Inn, dated 1535, an old, peaceful, low-ceilinged pub alongside a swing bridge which affords the customers sitting at the water's edge some interesting activity. The towpath here provides good, comfortable walking as the canal continues its climb out of London with several evenly placed locks as it approaches the small, handsome town of Berkhamsted. The town and the remains of the Norman castle can be approached from Bridge 141, where the road bridge has a footbridge originally used by barge horses crossing between the towing paths on opposite sides of the water.

The Berkhamsted School buildings, with the huge St Peter's Church beyond, can be seen from the bridge. John Incent, Dean of St Paul's, who died in 1645, founded the grammar school, which is of red brick with mullioned windows. Graham Greene, the author, was a pupil here. William Cowper, the poet, was born in the town in 1731, when his father was rector. St Peter's really is a colossal church, the length of half a football field and one of the largest in the Home Counties. Much is thirteenth- and fourteenth-century and, as you might expect, even though

Cottage at the entrance to Berkhamsted Castle

Cowper left the town when he was six years old, he has a memorial in the church, in the east window.

It is pleasant to wander from the canal down the street with its fine old houses, particularly Court House, which was built in the sixteenth century. The Canadian totem pole which stands in a timber yard close to the bridge looks a little out of place in this old and beautiful place. It was carved by a Vancouver Indian and put here in 1967.

On the other side of the bridge and just along the road is

Berkhamsted Castle, built in the days of William the Conqueror, who gave the manor to his half-brother Robert. There is not all that much to see of the castle today. Historians will revel in the magnificent earthworks and moats and soak up the surroundings but any little boy going to the castle would be disappointed at the ruin, with little to see except the vast expanse of grass, the remains of the walls, and the 45-foot-high mound with steps to the top. He would probably not even be impressed by the list of people who have lived and stayed here, among them kings and queens, the man who was murdered in Canterbury Cathedral, Thomas Becket, and the Black Prince, who is said to have regarded this as his favourite home.

I left the canal again at Bridge 139 and took the five-minute walk into Northchurch, whose George and Dragon is a coaching inn dating from the sixteenth century, with eighteenth-century additions, where gleaners took grain to be threshed in the yard. Another attraction, right opposite the George and Dragon, is St Mary's Church, getting on for a thousand years old and the oldest church around. In the porch behind the locked gate I saw the harvest offerings.

Outside the church door is a gravestone to 'Peter the Wild Boy'. He was found wandering in Hanover when he was thirteen, walking on his hands and feet like an animal, climbing trees and eating grass. King George I was instrumental in bringing him to Britain, where he became an object of interest at Court. George II's Queen Caroline even tried to educate him, but without success. His picture is on a brass inside the church and we are told that, 'After ablest masters had failed to make him speak he was sent to a farm, where he ended his inoffensive life in 1785, aged about 72 years.'

A disused stable and supply point for canal horses stands at the next bridge, and close by, a hundred yards downhill, is Forge Cottage, where the horses were shoed. Dudswell Locks are followed by the strange-sounding Cowroast Lock, close to the village of the same name standing on the A41 which has been close at hand for more than twelve miles. They say it should really be called Cowrest, because this was where cattle-drovers heading for London stopped for the night. There is a Cow Roast Inn as well, 150 yards from the canal on the main road, and a Cowroast Marina, mainly stocked with narrowboats when I walked by.

For the first time since leaving London, the canal enters a cutting as it approaches Tring railway station. The canal, which has been river-like in parts, is a true canal again, still and peaceful, and here is its summit. Anybody going the other way would find it downhill all the way to the Thames.

Aldbury is a mile to the right from the bridge near Tring station, rather a long way but well worth the effort for anybody with the time and energy. There is a timbered manor house, village pond, stocks and fine old thatched cottages, and a column nearly 200 feet high, erected in 1832 to the Duke of Bridgewater, the father of the canal system. To the left, but twice as far away, is Tring, close to Buckinghamshire, a small town of winding streets. The cutting, which provides lovely walking through a good collection of trees for about forty minutes, opens out close to Bridge 133 and the Grand Junction Arms. Here is the village of Bulbourne, which originated with the canal and heralded the branch to Wendover. The seven-mile Wendover arm was abandoned in 1904 but recent restoration has made it navigable to Little Tring – about a mile and a half – and walkable to Wendover.

I walked downhill with the locks, crossed the canal at the first bridge and back over at the second just after passing the six-mile arm to Aylesbury on the other side of the water. The Aylesbury branch remained a commercial proposition into the 1950s and, after deterioration, welcome restoration was started in the 1970s on a picturesque piece of canal.

The village of Marsworth was approaching, and its church could be seen on the hill to the right. It was late, but a thatch-roofed shop near Bridge 130, close to Marsworth, was still open. Pitstone, with its huge, grim cement works, came up on the right, and beyond it a windmill more than 350 years old and believed to be the oldest in the country. The Duke of Wellington pub sells good food and good beer, and over to the left is the small, undistinguished town of Cheddington.

This had been a superb stretch of canal, with perfect, tranquil, easy walking.

OS MAPS 166 and 165
Information Centres: Rickmansworth (0923 776611), Hemel Hempstead (0442 64451), Berkhamsted (0427 4545)

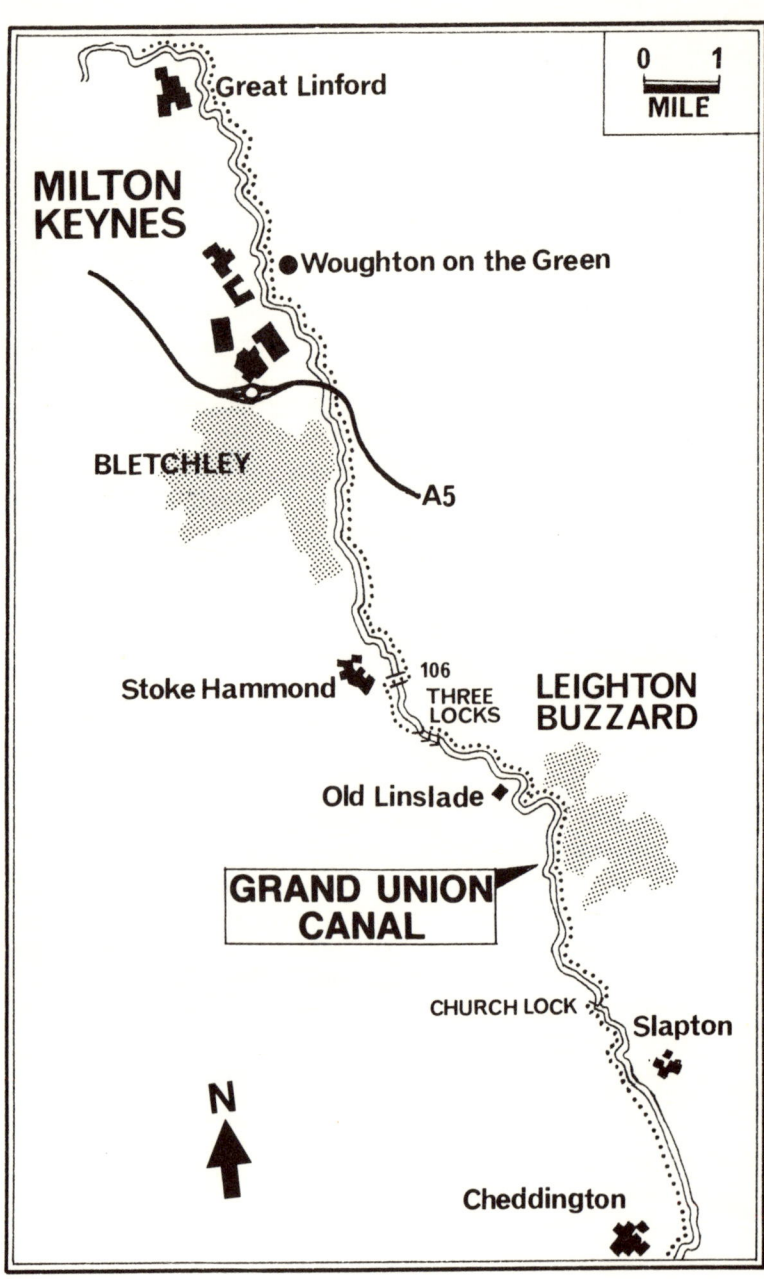

0   1
MILE

Great Linford

**MILTON
KEYNES**

●Woughton on the Green

A5

BLETCHLEY

Stoke Hammond

106
THREE
LOCKS

**LEIGHTON
BUZZARD**

Old Linslade

**GRAND UNION
CANAL**

CHURCH LOCK

Slapton

N

Cheddington

4

Cheddington to Great Linford

22 miles : 35.2 kilometres

The day was grey and once again not so promising. But days are like life, not to be taken seriously, certainly not at face value. The mist clung to the canal but, while it narrowed the view, it could not totally smother the sound. So while there was next to nothing in sight, the sound of the trains broke through the mist.

After a while, I found I could just make out Ivinghoe church to the right, before Ivinghoe Beacon popped up behind it. Despite the greyness I thought I could make out the Whipsnade Lion at the bottom of the hill to the right. But no. A man with two barking dogs on a boat told me I would see it 1½ miles further along the canal, round the other side of the beacon. The lion, which was carved out of the chalk in 1935, is over 480 feet long and should have been visible. But I never saw it, however hard I looked. It must have been hiding in the mist . . .

A heron floated out from the bank as the mist cleared and the country opened out with extensive views all around. I felt encouraged to have a look at the village of Slapton, a ten-minute walk away. But I need not have bothered. There was little to see and I was glad to get back to the canal and pure country with no sounds, except those of the birds and just the occasional faint rattle of a train in the distance. Several ducks flew off, and then a pair of herons picked themselves up to float away over the ploughed fields before leaving me alone to walk on to Church Lock.

The boats had begun to thin out but one arrived just as I reached Church Lock. Four middle-aged ladies were aboard the boat from Cowroast which scraped alongside the canal with a shout of 'Oh, damn' from somewhere in their midst. There were reassuring cries from the others: 'Never mind. You've made it.' It made me wonder what the rest of the journey had been like, how many other scrapes and scratches and head-on confrontations there had been. I decided to ignore them and concentrate, instead, on Church Lock House, once a fourteenth-century chapel, now a private house.

Five minutes further on came Grove Lock with, so I understood, an original Grand Junction milestone. It took me ages to find it. I wandered round the lock, on both sides of the water; I nosied about the lock house and searched in the shrubbery for the thing. A boat-load of people helped me as they dropped through the lock, but we could not find it. Then, after they had gone and I was about to give up, I came across it, a small mile-post, now painted in black with the words almost obliterated. I could make out 'Miles' and 'Thames' and had to take my handbook's word for it that it was '45 Miles to the Thames'. A piece of tramline on the canal bank a little further on is a reminder of the days when the goods were moved to and from the barges by waggon. It runs for twenty yards before sliding off into the fields where cattle now graze.

Leighton Buzzard, a lovely market town, was approaching. Through the trees I glimpsed the spire of All Saints Church, a building which dates from 1288. It holds a thirteenth-century font, fifteenth-century misericords and some interesting medieval inscriptions. Girls were playing hockey on the left – the first sign of organized life for hours – and the countryside reluctantly gave way to town soon after reaching the old railway bridge which formerly carried the Linslade to Dunstable line and which is now a public footpath. The biggest town since Hemel Hempstead, nearly twenty miles behind, had arrived, a pleasant, historic place only a few minutes walk from the canal.

The people who had helped me try to find the milestone at Grove Lock were preparing to leave their boat. They asked how far I was walking and when I told them to York, one lady exclaimed: 'What a lovely idea!'

A few minutes' walk took me into the market square, with the old fire station and the 600-year-old, five-sided market cross at the top. This really was worth the detour, a delightful town centre with many old buildings, including one, the Golden Bells Inn next to the Post Office, which is even older than the cross. The old fire station was housing stock for the market traders who go there on Tuesdays and Saturdays. It also houses the British Legion and

the WRVS community room, which is open most mornings for snacks and where I got a welcome cup of tea. I was surprised to have the place, with a selection of comfortable armchairs, to myself. 'We used to get a lot in but there was too much smoking,' the lady told me. 'It was awful. No ventilation, you see. We put up "No Smoking" signs, and now they've all gone.'

I was soon back by the canal, refreshed, and quickly enjoying the walk to Leighton Lock. Gardens, all clean and tidy, reach down to the water. The branches of trees hang into the water, and the Globe, low and whitewashed with tables outside, is situated on the edge of the water just before Bridge 111. There are deep marks, caused by the towing ropes, in the iron posts on the edge of the bridge, another reminder of days long and for ever gone. The countryside here is beautiful as the canal twists its way north, almost encircling Old Linslade church, which is in view for a long time. The River Ouzel wanders about on the right-hand side, and the scene is perfect . . . peaceful and perfect.

A milestone told me it was forty-three miles to Braunston, but who wanted information? The canal, its waters totally still, meanders its way through rolling country that contains little more than farms, trees and wild life.

Three miles after leaving Leighton Buzzard I arrived at the attractive area of Soulbury Three Locks with its craft shop, whitewashed buildings and pub. The pub, called Three Locks, contains canal mementoes, including beams from the original stables which housed the canal horses and which have been incorporated into the bar. It provided the first signs of human life for over an hour.

Then I returned to the country, stretching away on the right and looking as still as a photograph. Herons lifted from the banks, and the only sounds to disturb the utter quiet were those familiar little plops in the water and rustlings in the hedge. The village of Great Brickhill, with a church that is mostly 700 years old, showed up on the hill to the right before a decrepit, unused swing bridge, rapidly becoming overgrown, came into view.

Soulbury, with the lovely Liscombe House in a large park, a fourteenth-century church and a pond on the green, is a mile from the canal. But Stoke Hammond is a little nearer, another attractive village, this one with thatched cottages and a seventeenth-century inn. Two yew trees in the churchyard were

Leighton Buzzard, with its 600-year-old Market Cross in the
foreground and the old fire station behind it

planted in 1687, and the church is worth a visit for the canal traveller because of the twentieth-century window which shows the church, the canal bridge and a horse-drawn barge.

Herons were plentiful here but the country started to give way to more life as the overlapping towns of Bletchley and Fenny Stratford drew near. A road comes in close as Bletchley gathers strength, with traffic starting to take over from the more familiar farms, trees and fields. The right-hand side of the canal manages to continue to look rural, but urban life grows stronger with every minute on the left hand. The old village of Bletchley, with St Mary's Church and a few attractive old cottages, is too far away for a diversion. It stands on the far side of the railway junction which was once important and helped the town to grow. Fenny Stratford is that much closer, with a grand station building that is right out of the railway era, a gathering of houses and a pub close to the canal.

Not all that far away, where the railway crosses the A5, used to be the terminus of the London to Birmingham railway. For about a year, during the completion of Kilsby tunnel, the line operated in two halves. By 1837 trains from London were able to run to Denbigh Hall, where the line crossed Watling Street, then the passengers had to transfer to stage-coaches and horse buses, owned by the railway company, to get to Rugby, where they could resume their railway journey to Birmingham. Denbigh Hall was a small station with an hotel and had taken its name in an unlikely way, dating from the days long before railways – and canals, too – when the Earl of Denbigh was caught in a blizzard during a journey along Watling Street. He found shelter at a cottage, the home of Moll Norris, and stayed there until it was fit for him to resume his journey. When the Earl asked for a bill, he was not given one, and from then on the cottage became known locally as Denbigh Hall, handing on its name to the station and hotel which the company built at the same place. It did not stay there long, being pulled down in 1838 when Kilsby tunnel was opened and the line completed.

Another feature of Fenny Stratford is its lock with a drop of only twelve inches, created because the north section would not at first hold water. The building of the lock solved the problem.

The village of Simpson which lies between the River Ouzel and the canal has a small green with a thatched cottage but quite

One of the original Grand Junction Canal Company's
milestones

a large church, dating from the thirteenth and fourteenth
centuries. The towpath here is good and clean, with well-cut
hedges and a few seats.

A pair of tufted ducks took off in some alarm as a bearded man
on a bicycle passed by. I moved over. 'Thank you. Good afternoon. Lovely day,' he called out as he whizzed by. A play-school
was in noisy session at Bridge 91, and the next bridge surprised
me by bringing back the countryside.

But it was all a mirage. The fields soon vanished again, and
five minutes later I was witnessing builders at work, brick upon
brick upon brick, at the start of the Woughton Marina with
moorings for 125 boats. Woughton on the Green is just to the

right – 'Wuffton' is the pronunciation I am told – with its early eighteenth-century manor house, now an hotel with a big garden, close to the large green which is set back between trees.

Three more bridges on is an interesting housing development, and here are the Woolstones, Great Woolstone and Little Woolstone, never over-populated areas which were falling into decay until the development of Milton Keynes.

Your genuine, original village of Milton Keynes still exists to the right, about a mile away from Bridge 83. But the new town has taken over, created in 1967 and now with a population of over 100,000. The new housing designs are not everybody's cup of tea, but the little I saw from the canal I liked. And the developers have been happy to incorporate the canal in their plans, making the towpath walking here extremely pleasant, despite the closeness of the built-up areas.

Newlands Park runs alongside the canal after Bridge 82, and it was around here that I spotted the Buddhist Peace Pagoda, peeping over the trees, looking rather like a five-year-old's drawing of Mum, with a fishbone for a body and little arms and head.

The hamlet of Willen is less than a mile to the right before Pennylands Boat Basin arrives with its swish-looking houses with boats pulled up to the back doors.

Great Linford has been described as the most attractive village in the Milton Keynes New Town. Arthur Mee, in his book on Buckinghamshire, wrote of the 'scene of dreaming peace'. It is all true, with its thatched cottages, manor house, church and almshouses, taking you back in time with the utmost ease. I stayed the night here and enjoyed every minute of its peace. And to crown it all, there was the best pub I sampled, the Black Horse, a lovely, friendly pub alongside Bridge 76, serving good beer and delicious food.

It had been another wonderful day – long perhaps, but full of good walking and surprisingly pleasant surroundings, despite the arrival of such a big and modern town as Milton Keynes.

OS MAPS 165 and 152
Information Centre: Milton Keynes (0908 691995)

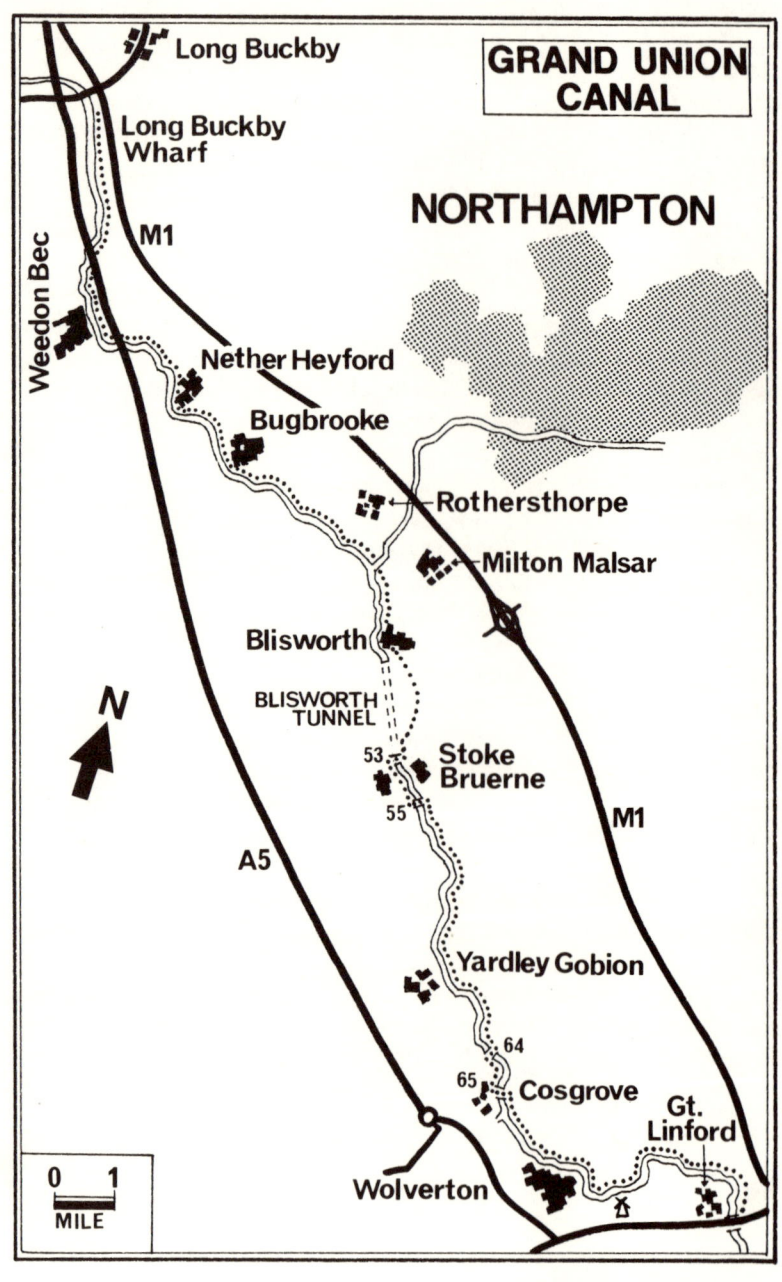

GRAND UNION
CANAL

NORTHAMPTON

Long Buckby

Long Buckby
Wharf

M1

Weedon Bec

Nether Heyford

Bugbrooke

Rothersthorpe

Milton Malsar

Blisworth

BLISWORTH
TUNNEL

N

53

Stoke
Bruerne

55

M1

A5

Yardley Gobion

64

65

Cosgrove

Gt.
Linford

Wolverton

0   1
MILE

5

Great Linford to Long Buckby

25 miles : 40 kilometres

A milestone soon informed me that Braunston was now only thirty-one miles away. I had come sixty-three miles from Brentford along the Grand Union.

The sun was still low, glistening and glaring in the water, and the sheep were tucking in to the grass as New Bradwell and its impressive windmill came near. I left the canal at Bridge 72, walked up the road for 150 yards, then down to the underpass and the path that led to the windmill, set back on the left. The windmill, built in 1816 and used for more than fifty years to grind corn, was restored in 1976. It stands in well-kept grounds and is worth the detour.

Lorries rumbled to and from a gravel works close to the canal before a huge black-and-white wall mural came into sight. It is all of 150 yards long, with a train engine at each end, and was painted in 1986. It gives the distance to Birmingham as seventy-two miles, Stoke Bruerne eight, Cosgrove two, Aylesbury thirty-two, and Brentford sixty-six, with eighty-one locks. A real train thundered by only ten yards from my ear before Wolverton, with its British Rail carriage works, slid past on the left. Wolverton was created for the works of the London & North-Western and produced locomotives and carriages. As the industry fell behind, a sign pointed along the canal to the Ouse Valley Park, leading ten minutes later to the great Ouse aqueduct, one of the major engineering achievements on the canal, with its history presented on an information board at the start of the bridge.

The lowest point in the canal between the two summits at Tring and Braunston is here at the River Ouse. Locks were used at first, nine of them in all, but they presented too many problems. The first aqueduct, too, proved troublesome before it opened in 1805. Three years later, on the night of 18 February 1808, it collapsed and the locks had to be brought back into use. The *Northampton Mercury* reported that Wolverton was 'entirely inundated to a great depth'. 'On repairing to the spot, however, it was found that one of the arches, which had been propped up underneath with timber, was still standing. This one arch, owing to there being no flood in the river, was able to carry off the water of the river as fast as it came down.'

The brick aqueduct was replaced with a cast-iron trough, a relatively new and advanced method of spanning a valley. Benjamin Bevan, an engineer from Leighton Buzzard who worked for the canal company, designed it, and it has remained solid and unmoved since 21 January 1811. In fact, there have been only two stoppages for maintenance work, in 1921 and 1986. There are passages through the embankment under the canal – 'cattle-creeps' – to give cattle access to the fields on either side of the canal. One of the aqueduct pillars carries two dates – 1811 and 1921.

Cosgrove marks the start of a long stretch through North-amptonshire, tucked close to the county boundary marked by the River Ouse. The canal passes through Cosgrove Lock, the first lock since Fenny Stratford twelve miles back, and immediately hits the Stratford and Buckingham Canal on the left. Little remains of this branch, only 150 yards still being in water and providing moorings. When it was opened in 1801, it was 10½ miles long with two locks and went via Old Stratford, Deanshanger and Thornton. Main traffic was hay and straw for London, and a boat-building yard was created, with steam launches, tugs and riverboats being built there for more than eighty years, until after the First World War. Work on the A5, which passes through Old Stratford, caused the abandonment of the length to Buckingham in 1961, and the rest went when the bridges were lowered. But the route can still be traced and a nature trail runs by the canal.

The Georgian house close to the lock is Cosgrove Hall, and a feature of the village is the strange little tunnel under the canal which links up with the Barley Mow. There is not much room – I had to bend to get through – but for giants who are out walking there is the nearby Solomon's Bridge, number 65, built in 1800 in the Gothic style and quite splendid. The towpath, which had been unbelievably good ever since London, carries a sign here boasting 'Towpath Improvement, sponsored by Northamptonshire County Council as part of Manpower Services Commission Community programme, April 1985'.

The road to Northampton was to stay around for a while but the next few hours were mostly in the country. It was back to the fields and the quiet, the peace of England.

The last of the bridges with a Milton Keynes number was behind me. They seemed to have been around for a long time, the MK label accompanying the number. The village of Castlethorpe is about a mile to the right from Thrupp Wharf, and it was soon after here, from around Bridge 63, that I encountered the worst piece of towpath since leaving London, seventy miles behind. The countryside was perfect, but the towpath was overgrown and unpleasant. There was no slowing down or fighting with the hedges, but for a mile or two I had to brush through the overgrowth, thankfully being cut back from Bridge 62 by two men.

'The path's not bad your way, not after what you've come through,' said one.

Yardley Gobion shows up on a hill to the left, a pleasant-looking village, soon followed by Grafton Regis, which was the scene of royal intrigue and love more than 500 years ago. King Edward IV went there secretly to marry Elizabeth Woodville, the widow of a Lancastrian knight, in the chapel of her father's house at Grafton Regis, the only witnesses being her mother, two gentlemen and 'a young man to help the priest sing'. When the Queen's father was raised from a baron to an earl, and her family married into the wealthy families of the land, there was a lot of ill-feeling, and there was a rising in Yorkshire to end the influence of her relations. The rebels set off for Grafton Regis, and the opposing armies clashed at Edgecote, fifteen miles away, where the royal army was defeated. The Queen's father and brother were murdered. There is a manor house still at Grafton Regis, and a small part of it is from the original house which stood here 500 years ago.

The path was still a little overgrown, but an adequate single-file track through the undergrowth had been made by anglers. The sound of traffic to and from Northampton can be heard on the left, and to the right is the River Tove on its way to join the Ouse at Cosgrove. Just ahead is the start of the seven-lock system that lifts the canal to Stoke Bruerne, one of the joys of canal-lovers.

The windmill at New Bradwell, built in 1816 and restored in 1976

Warehouses and cottages, a pub and a tea-room stand proudly at the edge of the canal at the village of Stoke Bruerne, side by side like troops ready for inspection, and they are immaculate. It seems that a man called Jack James once lovingly tended the locks and displayed traditional canal ware in the old leggers' hut, winning awards for his efforts, and in 1961 it was decided that Stoke Bruerne should become the site of Britain's first canal museum. It has seen well over a million visitors since it opened in 1963.

A mile-post which declares 'Etruria 16 miles Uttoxeter 14 miles' reminded me of the day three years earlier when I walked the Caldon Canal from Froghall to Etruria in the Potteries. There was a winch that had operated the locks on another canal I knew well, the Fossdyke and Witham, and a parapet lamp from the bottom lock of the Wolverhampton flight of the Birmingham Canal Navigation. Among the buildings is a cottage offering bed and breakfast, a welcome sight for the walker and only the second I had seen in seventy-four miles from Brentford. The other, which had been advertised at Yardley Wharf only a few miles earlier, had been for farm accommodation. Close at hand is Stoke Park, with its beautiful gardens which are open to the public at weekends and Bank Holidays in the summer.

As I prepared to leave Stoke Bruerne, doubly refreshed at the tea-room and the Boat Inn, so did a boatload of youngsters heading for the mysterious darkness of Blisworth Tunnel. A sign proclaimed: 'Following extensive repair work, re-opened 1984.'

Blisworth Tunnel was the last link in the Grand Junction and was opened in 1805, five years after the rest of the canal had been completed. In those five years boats were unloaded onto horse-drawn waggons which were then pulled over the hill along a tramway and re-loaded on to the boats. The tunnel, 1¾ miles long, is second only to Dudley Tunnel – by ninety-eight yards – as the longest still navigable in the country. A large amount of re-lining was done in 1984 at a cost of £4½ million, and maybe if I had been in the boat with the youngsters I would have been able to see evidence of it, but I was not and, as there was no towpath, I had to go over the hill, following the lane and road to Blisworth. The road hugs the line of the canal, and six air shafts are a further reminder of its position, one of them at Buttermilk Farm. After half-an-hour's walking through rural Northamptonshire I heard a boat chugging and youngsters chattering and saw below me the boat I had seen leaving Stoke Bruerne. I regained the canal by following the road through Blisworth, a good-looking town with a dust-covered main street running between Oxford and Northampton.

The canal now makes its way to Weedon Bec, a small but important town standing at the crossroads of the A5 (Watling Street) and the A45 from Northampton to Coventry. It wriggles its way past the village of Milton Malsor, an attractive, dreamy spot with its fourteenth-century church, houses of stone and thatch, and tidy gardens, and which stands even closer to the five-mile-long Northampton branch of the Grand Union. Then come Gayton, on a hill with several large, striking old houses; Rothersthorpe, facing Milton Malsor across the Northampton arm; Bugbrooke, and Nether Heyford, where a school was founded in 1673 for the children of Nether and Upper Heyford, and for all those children called Bliss living within five miles.

The towpath here was hard going. The hedges closed in; I fought and argued with them but could not avoid them, and progress slowed as I battled my way through. I could see little beyond the hedge but the noise told me the M6 motorway was coming near.

St Werburgh, daughter of a King of Mercia in around the seventh century, founded a convent at Weedon; a Royal Military Depot was established there in 1803; stage-coaches between London and Holyhead went this way, and today it is still a busy junction supporting several hotels, with the Crossroads, the Globe, the Wheatsheaf and, near Bridge 26 on the outskirts, the Narrow Boat. The remains of the barracks still exist – they were situated here during George III's reign in case of a French invasion, because Weedon was, they reckoned, furthest from any of England's coasts. The railway line from London runs through here, too, and in the days when it had a station Queen Victoria alighted here in November 1844, when she and Prince Albert were going to Burghley House, Stamford, to attend a christening, travelling from London Euston to Northamptonshire by train. Weedon was the most convenient station, although the royal couple still had forty-four miles to travel by coach.

The country opens out beyond Weedon, although the railway, A5 and M6 are close at hand, so close at one point that little more

than 500 yards separates them, the canal and the River Nene! The
towpath, which became good again just before Weedon, keeps
close company with the motorway around Whilton Locks, and at
one stage I reckoned I was less than a cricket-pitch length from
the road.

It was here, on the Whilton Locks section, where the accommodation I had expected to find turned out to be non-existent. I
was told that nobody in the area provided bed and breakfast.
I was desperate and weary, and the day was turning dark as I
carried on up the canal. Within minutes I saw the lights of a shop,
Anchor Cottage Crafts, by the towpath and I was soon pouring
out my tale of woe. Without hesitation, Brenda Walker offered
me a bed, that of her son who had recently returned to university.
She and her husband Philip welcomed me into their home, and
the following morning I was sent on my way with a wonderful
breakfast and a large bunch of grapes.

OS MAP 152
Information Centres: Milton Keynes (0908 691995), Northampton (0604
22677), Rugby (0788 2687)

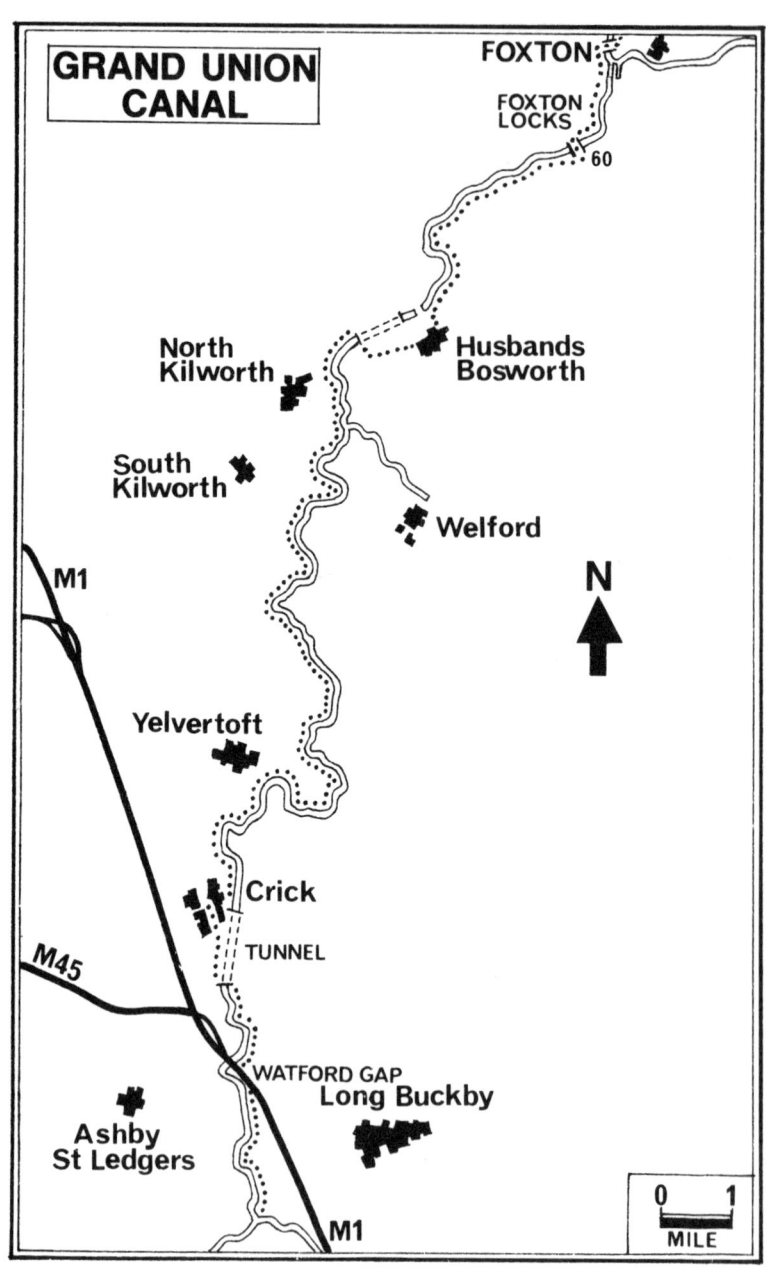

GRAND UNION CANAL

FOXTON

FOXTON
LOCKS

60

North
Kilworth

Husbands
Bosworth

South
Kilworth

Welford

N

M1

Yelvertoft

Crick

TUNNEL

M45

WATFORD GAP

Long Buckby

Ashby
St Ledgers

M1

0   1
MILE

6

Long Buckby to Foxton

23 miles: 36.8 kilometres

I wakened to the comforting quiet of deep country. From my bedroom window I looked out on the lock and a deserted scene and thought of settling down there for a week or two. Unfortunately, I had to leave.

Long Buckby, which is about a mile and a half away, was a worsted weaving centre before turning over to that familiar Northamptonshire occupation, shoemaking. And it was from here, in 1953, that the Household Cavalry ordered 275 pairs of jackboots for the Queen's Coronation. They were made by local hand-sewers and were worn by the Sovereign's Escort. Some of the same workers had made jackboots for the Coronation procession of King George VI sixteen years earlier.

There is a story about Long Buckby Brass Band who played for dancing one day at a Syresham Friendly Society event. The custom was for dancing to go on in the school yard all afternoon and evening after a church service, but on this occasion the band members fell out among themselves about one of their number who was not keeping time. They abused one another, began to fight and then turned their wrath on one another's instruments. The big drum was kicked in, the cornets and trombones were smashed and the remains were thrown over the school yard.

Just up from Long Buckby Wharf, across the busy A5 trunk road, is Norton Junction where the canal splits in two. The main line goes off to the left through Braunston Tunnel, over a mile and a quarter long, and to the canal village of Braunston, before running off to Birmingham. The other line was mine, heading for Leicester and its rendezvous with the River Trent just below Nottingham.

A sign, declaring Leicester to be 41½ miles away, stands close to a field where potatoes were being picked by machine. I passed under the A5 again and the sounds of rail and road came close as I walked under the railway line – the old LMS – beneath a long, gloomy 120-yard bridge which brought me out right alongside the M1 motorway. This was one of those culture shocks you do not expect on the canal bank after miles of quiet country walking, coming almost literally face to face with the mad, mad world of the motorway. I was right next to Watford Gap service station, just a fence away, and if I had not just set out I might have been tempted to climb the fence and head for the service station and coffee. Lorries and cars, their headlamps blazing through the morning mist, pulled in to the car-park. I watched them for a few minutes, found nothing to interest me, then set off again. After all, I had come here to avoid all that.

The towpath, which had been difficult but manageable, came good again near the Stags Head at Bridge 5, a few minutes before Watford bottom lock came into view. The mist and the trees restricted the view but nothing could yet deny the noise of the railway and the motorway, both of which were thankfully taking themselves off into the deep grey yonder.

There is a footpath from the locks to the village of Ashby St Ledgers to the left, a beautiful, clean village which is a mile and a half away but worth seeing. I first came on it by chance some years earlier and have been drawn back to it since, particularly to its pub, where a goat once came in and stood on its hind legs at the bar. Ashby St Ledgers was the home of the Catesbys, one of whom, William, who supported Richard III, was beheaded after the Battle of Bosworth. Yet strangely his tombstone in the church gives the date of his death as two days before the battle. The Catesbys were involved in an infamous piece of English history more than a century later, when Robert Catesby became involved in the Gunpowder Plot. When Guy Fawkes was arrested, Catesby and his friends galloped the eighty miles from London to Ashby and then on towards Wales. They were cut off at Holbeach, and Catesby was shot dead. The half-timbered gatehouse, where the plot is said to have been hatched, is still there.

If Ashby is a little too far away for comfort, there are the Watford Locks to consider, a seven-lock flight with four of them forming a staircase where the top gate of one lock is the bottom one of another. For the first time since leaving Westminster I found myself confronted with a real hill to climb as the locks lifted the canal 52½ feet in a short distance.

There are no more locks now till Foxton twenty miles away, the summit of the canal, with all flat and downhill between there and the River Trent. The path deteriorated again after the locks, and it was something of a relief to leave the canal at Crick Tunnel, getting on for a mile long. I swung off to the left along the old horse-path, reaching the top and going left on the lane and road to Crick, which is signposted. Boat Horse Lane on the right invited me back to the canal but I decided to go on to the village with its thatched buildings, church spire and Crick Manor. The large church of St Mary contains a 'sentry box' which was used to protect the priest from the rain during funerals.

There are no villages on the canal on this stretch, so Crick and nearby Yelvertoft are handy places to call at if in need of refreshment. Yelvertoft is reasonably near Bridge 19, an interesting place where Percy Pilcher, an aeronaut, was killed in the park in 1899, four years before the Wright brothers first appeared. Pilcher, who was thirty-three, was flying a gliding machine at the time of his death but was already designing a small petrol motor to power a larger model. There is a monument to him in the park by the River Avon.

The canal moves off into the depths of the country, wandering like a drunk, back and forth as if uncertain which way to go. But it was a lovely day, with countryside to match, and the walking was perfect. Then, quite suddenly, the towpath, which had been behaving itself beautifully, turned nasty near Chester's Bridge. I was forced to push my way through the brambles and, if I was not scratched, I was sure to trip over them. This was the worst section of canal I had encountered in nearly a hundred miles walking, with brambles, hedges, weeds and chest-high nettles all taking their toll. I watched and listened desperately for a boat, so I could get a lift until the path improved, but there was no sign. Just when I needed help, there was none forthcoming, and all I could do was fight my way through a dreadful section of path which took away much of the enjoyment of the day. The poor walking persisted, so I escaped into Welford, a large red-brick village on the Leicestershire boundary.

I had come here some years earlier to walk the Welford arm of the Grand Union, completed in 1814 and only a mile and a half

The downhill run on the two pairs of five-rise staircase locks at Foxton

long. This towpath, too, was untended but the going was a little easier along a lovely little piece of the canal system that was forgotten for too many years before being restored and reopened in 1969. It was clear the wildlife was not used to being disturbed. There were all sorts of rustlings in the hedges, and birds were breaking out of the trees all along. Sheep and fences got in the way and I was surprised when a boat from Burton chugged its way up the branch. A bridge linked up with a farm close by, and a few cows, who had added their own squelch to the soggy path, watched me suspiciously. An old, decrepit wooden footbridge had no steps on my side and looked to have had a gate in the middle at one time. A lock with a single paddle and a bridge which takes the path to the other side come near the canal's end. This is a real backwater creek, ending at the wharf at the bottom of the village of Welford, hiding from sight behind the Leicester–Northampton road and providing boats for hire.

The main line of the canal continues north, bypassing North Kilworth on the left before swinging east and into Husbands Bosworth tunnel, nearly three-quarters of a mile long and with heavily wooded approaches. Husbands Bosworth itself is a pleasant town, so named to distinguish it from Market Bosworth. One was the market town, the other the town of farmers, 'husbandmen', and noted for its witch-hunts in the seventeenth century. In 1666 nine old women were murdered here, 'hostages to Satan' it was said, after a boy was not cured of epilepsy.

The horse-path follows – as near as does not matter – the line of the canal as it climbs over the hill, and comes out at the bottom of the town. The Northamptonshire–Leicestershire border is crossed as the canal goes over the River Avon just before the branch off to Welford.

I had been led to believe by my trusty guide-book that, once in Leicestershire, the towpath would be excellent. No such luck. The path was awful, terrible and horrible – in that order – for at least three miles more before suddenly becoming acceptable again within a mile of the Foxton flight of locks. There had been work on the path. The grass and hedges had been cut back and down, and the lovely Leicestershire countryside was there to be seen once more, with cows and sheep and farms and miles of rolling land.

The descent at Foxton is so steep that, as I approached, the

At the bottom of the Foxton Locks where the branch leaves for
Market Harborough

arms of the top lock seemed to reach out in apparent solitude. The rest of the locks are there to be seen from the top of the hill, two pairs of five-rise locks, as straight as a stepladder. It can take an hour to an hour and a half for a boat to negotiate the locks, and there can be bottlenecks and much longer delays on a busy day. In an effort to cut out the hold-ups, the Foxton inclined plane was opened in 1900. The plan was to float boats into tanks which then ran on rails up and down the hill. It was steam-driven, which proved costly, and the rails were constantly under stress from the weight they had to bear. The locks, looking today exactly as they did when they were opened in 1812, were re-opened in 1908, and the plane was closed in 1911. This is a lovely place to dally in, to marvel at one of the triumphant parts of the canal age of the eighteenth century. And if you are really lucky, the pub might be open too.

The branch canal to Market Harborough leaves Foxton at the foot of the locks, almost six miles of quite perfect walking to one of England's fairest market towns. It would put twelve miles on the journey but it is worth it if Market Harborough can be planned as one of the overnight stops. It is barely 2½ miles from Foxton as the crow flies, but the canal does a great arc round Gallows Hill as it wanders like a river and covers more than twice the distance. The village of Foxton stands on both sides of the canal, and around these parts John of Gaunt used to hunt stag when he was lord of the manor.

This had been a frustrating day in some ways, because, while the canal passed through delightful countryside, I had found myself battling too often with overgrown hedges to take full advantage of the sights. It had been hard going and proved to have been quite the worst walking day of the entire walk from London to York. It had been one of the loneliest days as well, as the canal dribbled its way between the villages, hardly touching any of them.

OS MAPS 152, 140, and 141
Information Centres: Northampton (0604 22677), Rugby (0788 2687), Market Harborough (0858 62649/62699), Leicester (0533 556699)

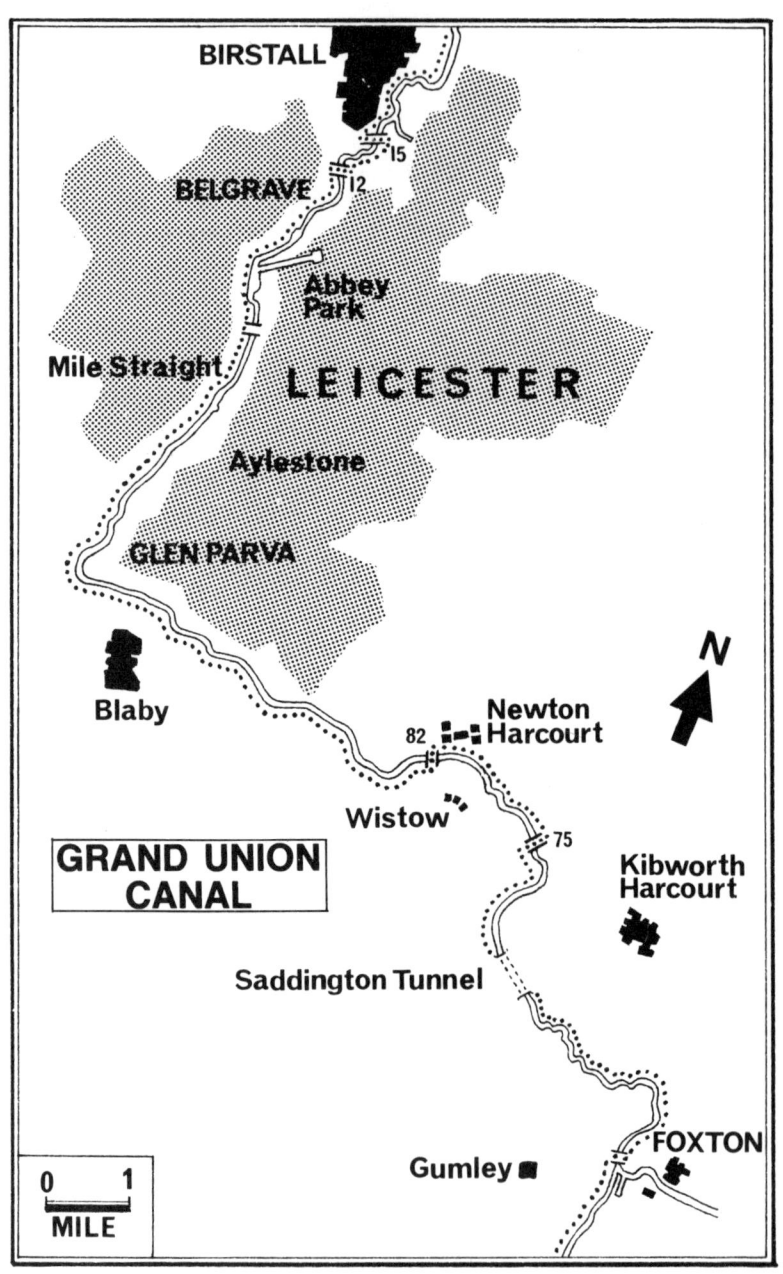

BIRSTALL

BELGRAVE 15

12

Abbey Park

Mile Straight

LEICESTER

Aylestone

GLEN PARVA

Blaby

Newton Harcourt

82

Wistow

75

Kibworth Harcourt

GRAND UNION CANAL

Saddington Tunnel

FOXTON

Gumley

N

0 1
MILE

7

Foxton to Birstall

22 miles : 35.2 kilometres

I was particularly looking forward to this part of the walk as it took in Leicester, the first city I had seen since London. Not that cities are usually all that special. It is the countryside that makes it all so appealing. But I had been here before and I knew that the walk through Leicester is rather distinctive and unexpectedly pleasant.

But first of all I had sixteen miles to go through the Leicestershire countryside, through fox-hunting country, with hardly a village in sight. Gumley, described by Arthur Mee as 'perhaps the prettiest village in the county, set amid steep hills and dales, with a footpath leading through a wood to its church and hall', is left on one side and the fairly featureless Smeeton Westerby on the other. Kibworth is over to the east too, sitting astride the great A6 trunk road and consisting of two villages, Kibworth Beauchamp and Kibworth Harcourt. Pleasing houses are the feature of both villages but Kibworth Harcourt is specially blessed with a lovely house dating from 1678 where Dr Philip Doddridge lived in the eighteenth century. The good doctor was a Nonconformist and a hymn-writer who was a pupil and teacher at the academy here. He enjoyed his life at Kibworth Harcourt. 'One day passeth away after another and I only know that it passeth pleasantly for me,' he once wrote in a letter.

Between them come Debworth Wharf and the half-mile-long Saddington Tunnel where the headless ghost of a woman decapitated in the tunnel is said to wander. Happily for the walker, therefore, there is no towpath but a path across the hill onto a lane with a hedge either side. The locks at Kibworth are followed by the aqueduct over the River Sence which trickles contentedly below, able to use only one of the two arches provided. The Sence stays with the canal for the next five miles before joining the River

Soar which is to become the Navigation and is canalized for most of the way to the River Trent, some thirty-five miles away. After the horrors of the previous day it was good to be on a well-maintained footpath in quiet countryside – quiet except for a yappy Yorkie called Sooky which shouted at me from a boat.

Wistow church stands in the fields to the left, with Wistow Hall in the trees just beyond, a Jacobean hall by a lake. The 800-year-old church was dedicated to Wistan, the Christian prince of the Saxon kingdom of Mercia who is said to have been murdered where the church is standing. The hall was the home of the Halfords for generations, and several members of the family are remembered in the church, including Sir Richard who supported King Charles I and gave him shelter before the Battle of Naseby. Sir Richard and his son Andrew were both sentenced to death by the Parliamentarians but were freed after paying ransoms.

A lane links Wistow with the village of Newton Harcourt, right by the canal, with an attractive Elizabethan manor house and a tiny thirteenth-century church. In the graveyard, close to the gate, is a monument to a child, in the shape of a church with a spire, windows and doors. The inscription reads: 'To the fragrant memory of Christopher V. Gardner, born 26 August 1916, died 20 September 1924, aged eight years.'

The towpath, which has been on the right-hand side of the canal for three miles from Bridge 75 due to the insistence of the Countess of Denbigh, moves back across at Bridge 82 near Turnover Lock. The Countess had required, before work started on the canal, that the towpath be on the side of the water opposite to Wistow Hall.

One of the features of the canals is the variety of lock names, several of them intriguing and lost in history. Around here are Bumble Bee Lock and Double Rail Lock, which was so named because the bottom gates at one time had two sets of handrails, installed after a woman fell in and drowned while crossing the lock. Those gates, it seems, still form part of a public footpath from Countesthorpe to Wigston.

South Wigston, with the first real housing and industry for many miles, was coming up on the right and a church spire peeped at me from across the fields. Leicester, I learned from one of the mile-posts, is only seven miles away, but the rural aspect persists on the left, although urban life is starting to dominate on

The thirteenth-century church at Newton Harcourt. In the
graveyard is a monument to a child in the shape of a church

the right. There are sidings and wharves at Blaby and Glen Parva, the railway stays close at hand for a few miles, yet this was still good walking with pleasant views and alongside a clean water.

A Beefeater pub, the fifty-year-old County Arms, stands alongside the canal at Bridge 98 as the canal's run alternates through country and industry, providing variety to the walk. Soon after passing Glen Parva the canal does a sharp right-hand turn, heading straight for Leicester after running alongside the city's southern boundary. The River Soar, which starts life around Wolvey some fifteen miles to the south-west, appears on the left, providing more glimpses of open countryside before life returns with a bang – an Asda supermarket and a gasometer.

King's Lock was about a mile away, and a sign told me that, 'The canal route through Leicester was completed in 1797 with the opening south from West Bridge of the Leicestershire and Northamptonshire Union Canal, although through traffic to London was not possible until 1814.' King's Lock heralds the nearby medieval pack-horse bridge at Aylestone which Charles I used after his flight from Naseby. He stayed at Aylestone Manor House during the siege of Leicester. The spire I saw from the canal was that of Aylestone parish church, with its splendid fourteenth-century chancel and Jacobean poor-box.

It is around here that the canal and the River Soar join company. The towpath through Leicester is part of a Riverside Walks scheme, and Leicester Riverside Park signs are everywhere, mingling with others to Aylestone Meadows Nature Trail. A parks policeman came by on his motor-bike, a man I was to see regularly for the next three hours as he patrolled the canal bank, from the south of the city to the north.

I walked between gasometers before reaching St Mary's Mills Lock, where the country is still fighting off the approaching city, a pretty place despite the sight of the gasometer looking over a factory and electricity pylons. Another sign points out the twelve-arch railway viaduct directly ahead, while to the left are St Mary's Mills and on the opposite bank Aylestone Gas Works. Then comes a huge expanse of water, with Freeman's Meadow Lock and the enormously broad weir, all rather unnerving I should think for inexperienced boat-handlers, who are warned to keep well to the towpath in this area. There are no such horrors for the walker, who passes, just through the lock, a stone which shows the level of the floodwater in 1912. Across the water can be seen Leicester City football ground, and straight ahead the start of what is known as the Mile Straight, a self-explanatory part of the canal and a perfect piece of walking, with a *real* footpath right through the heart of Leicester.

When L. T. C. Rolt travelled along the waterways of the Midlands in his narrowboat, *Cressy*, in 1939, he found little about Leicester to delight him. In fact, he was driven to say, as he sailed away from the city, 'We had seen enough of Leicester.' He had first of all travelled into the city from Aylestone by tram and seen the market and the church of St Mary de Castro with its lovely spire that today dominates the view of the canal traveller. The following morning he left Aylestone and headed for the open country, for peace and quiet after the noise and smells of the city.

The River Soar is Leicester's back door [he wrote], and as back doors are apt to do, it reveals 'domestic offices' which usually remain discreetly hidden from the eyes of visitors. Broad squares and pretentious public buildings proclaim the city's commercial prosperity to the traveller by road, but the water-borne traveller sees a very different picture. This is no less than the ugliness and squalor which underlie the superficial pomp and circumstance of all great cities. We saw the reeking gas-works, mountainous refuse dumps, the power-station with its gigantic steam-capped cooling towers, great mills pulsating with machinery rising sheer from the water's edge and, above all, the countless mean streets where dwelt the servants of these monsters.

I wished Mr Rolt could have seen it now. I wished he was here, fifty years on, to witness the transformation. Of course, there is industry, but Leicester has at least grown proud of its canal heritage and turned the entry into its city into one of the finest among Britain's large towns. It is a pleasure to walk through Leicester, on a path that is clean and firm, by water almost totally clear of rubbish and wide enough for at least two lanes of waterway – each way. Trees and gardens and parkland make for relaxed walking, and as I walked under a succession of ornamental bridges I, unlike Mr Rolt, was very happy to be in Leicester. The bridges are lovely, particularly West Bridge,

Gargoyles look out over the Grand Union Canal from West
Bridge in Leicester

opened in 1891, and I took time at every one of them to leave the canalside and stand and stare from them. I watched people hurrying by, scurrying like beetles, many of them seemingly unaware or oblivious of the historic, quiet waterway beneath them.

Leicester has remains from Roman times; there is a Saxon church, St Nicholas, the original hall of the Norman castle as well as the Norman church of St Mary de Castro, and many other medieval buildings. The Guildhall was begun in the late four-teenth century, and two fifteenth-century gateways survive as well as numerous buildings from Tudor times to the present day. Leicester exploded in population in the early nineteenth century – there were only about 17,000 inhabitants in 1800 – and is now one of the Midlands' most splendid cities. Other large towns and cities might have more tourist appeal but Leicester, as I have discovered through years of visiting it for county cricket matches, has so much to offer that two or three days are needed for exploration. And for the walker it is well worth making an overnight stop here.

The Mile Straight, which hosts the annual rowing regatta, is a memorable introduction to the city proper, opened in 1890 as part of a major flood-prevention scheme and leaving the river over to the left. From Walnut Street bridge there is the peculiar sight of a Statue of Liberty, rather smaller than the American one, of course, standing on the corner of a nearby building. Swimming in the river was popular in the last century but public sensibilities were aroused as a result of bathers 'wantonly exposing their persons'. Perhaps the existing notice on Walnut Street bridge to the effect that the water is unsuitable for bathing has something to do with an underlying fear that maybe the wanton exposure might start again . . .

Gargoyles look out over the canal from their vantage-points on West Bridge, probably the best bridge from which to enter the city, and the start of the old Leicester Navigation, opened in 1794. Evan's Weir is just beyond, and here was the site of a wharf where coal from the Swannington area was transferred from railway waggons to canal barges for shipment to London. Swannington, about twelve miles north-west of Leicester, gave its name to the Leicester and Swannington Railway, the first line laid south of Lancashire. Robert Stephenson played a major part in laying the

Friars Mill at Leicester, more than 250 years old

line, and his father, George, drove the first train on it in 1832. The canal system played its part in the historic day, having enabled the famous engine *The Comet* to be taken to Leicester by boat.

On the opposite side of the water to the site of the coal wharf stands Friars Mill, dating from 1735, and a little further north is a hundred-yard-long footbridge with five stone supports across the weir. Another bridge, not much shorter, stands at Hitchcock's Weir as the canal heads out of Leicester on the sixteen-mile run to Loughborough. For those with a head for figures a sign on the wall at Limekiln Lock tells the distances by waterway to various parts of the country. Behind us are Northampton 57, Oxford 100 and London 136, while to the side is Birmingham 78. Ahead are Nottingham 34, Lincoln 90, Manchester 145 and York 158, which gives the total distance from London to York as 294 miles, a little below my own estimate. But then, the canal does not go wandering off into nearby villages or have to find refreshment.

Abbey Park is close by, a lovely park with its flowerbeds and lawns, boating lake and shady avenue, bordered on the far side by the River Soar and containing the ruins of Leicester Abbey where Cardinal Wolsey died in 1530. Some of Leicester's famous hosiery mills stand by the canal, the most famous perhaps being the Wolsey Works which comes just before Swan's Nest footbridge, which took me to the other side of the canal for a few minutes and to Belgrave Lock where the canal rejoins the River Soar.

This had been an interesting, delightful part of the canal and provided me with one of my outstanding memories. I could not help thinking again of Mr Rolt and his miserable journey along these same waters fifty years earlier. What had he said?

> The water here was black and foul, the surroundings depressing in the extreme, so that we were glad when, after passing the vast Wolsey Underwear Mills, we sighted the river again and the first glimpse of open country. We had seen enough of Leicester. Urbs had one more card to play, however, for we found that Belgrave Lock, through which we had to pass to rejoin the river, was full of dead rats. The churning of our screw set their bloated, putrefying bodies bobbing up and down in horrid semblance of life. Had there been a local rat week? we wondered. Two imperturbable fishermen, who sat on the lockside gazing raptly at the water, discountenanced the theory, for they appeared to accept the malodorous corpses as part of the normal order of things. Perhaps Leicester breeds a fortuitous indifference to the more unpleasant aspects of city life.

Oh, how dearly I wished Mr Rolt could have been with me to see the transformation! He would have loved it.

Holden Street bridge, a cable-stayed footbridge opened in 1985 which replaced an earlier bridge over the Soar, is followed by the attractive Belgrave Gardens. Once a country village, Belgrave witnessed the growth of hosiery manufacture, industry and trade which by 1800 had become more important to the village economy than agriculture. In 1851 almost all the working population was employed in framework knitting, which was carried out mostly in the operatives' homes. The area between Leicester and Belgrave was built up with streets of terraced houses and factories in the last thirty years of the last century, and the village was incorporated into the city boundary in 1892. Belgrave is all part of the spread of Leicester, on the edges, to be sure, but still part of city life.

By now the canal has become a true river and, after passing a seven-arch bridge with a separate footbridge, it reaches for its more familiar rural surroundings again. The city had been lovely, really it had, but it was time for the country again, alongside a gently meandering river with Riverside Park posts directing me to Watermead Park. For the first time in about 140 miles I was no longer with a canal but with a river, walking through fields with stiles and revelling in the open spaces after the confines of the canal and the city.

OS MAPS 141 and 140
Information Centres: Market Harborough (0858 62649/62699), Leicester (0533 556699)

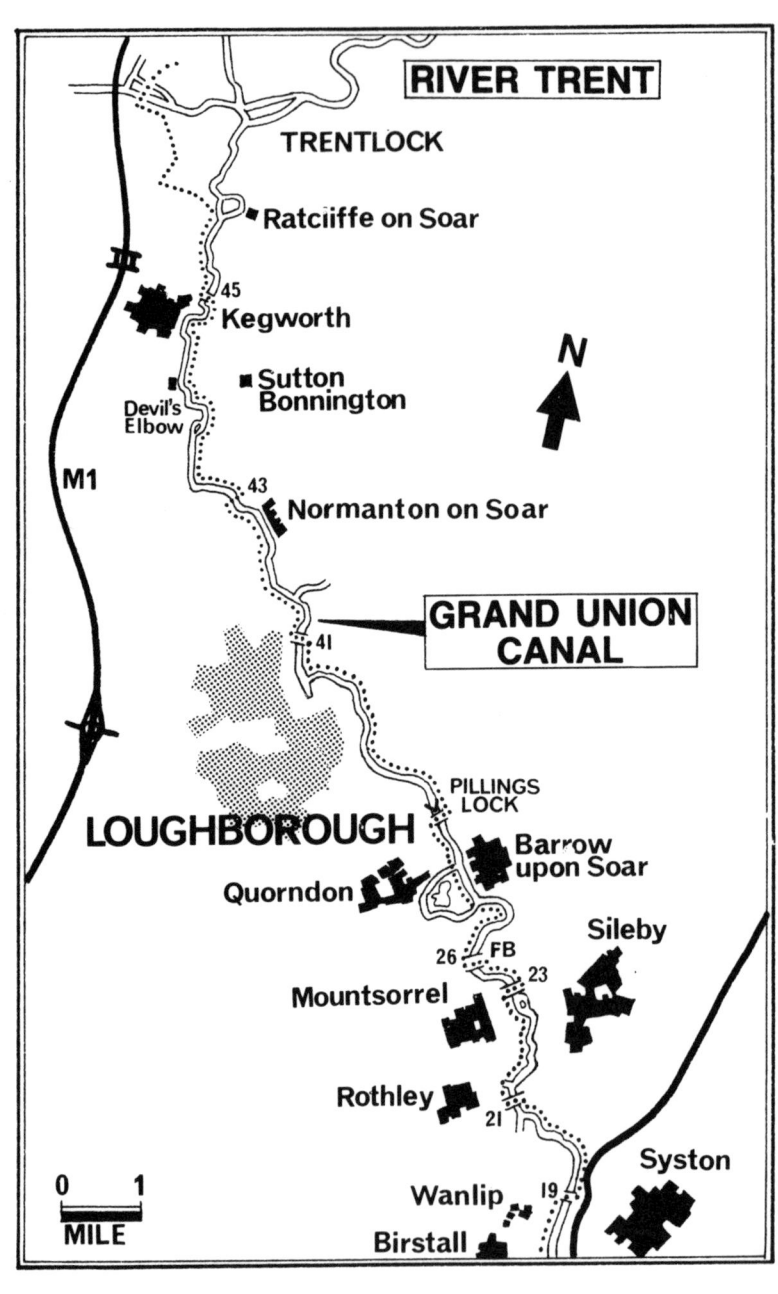

RIVER TRENT

TRENTLOCK

■ Ratcliffe on Soar

45
Kegworth

■ Sutton
Bonnington

Devil's
Elbow

M1

43
Normanton on Soar

GRAND UNION
CANAL

41

PILLINGS
LOCK

LOUGHBOROUGH

Barrow
upon Soar

Quorndon

Sileby

26 FB
23

Mountsorrel

Rothley

21

Syston

19

Wanlip

Birstall

N

0   1
MILE

8

Birstall to Trentlock

22 miles : 35.2 kilometres

This was the day that would take me into Nottinghamshire and the first real signs of the North. Nearly all the way would be by the River Soar, twisting and turning through fields, and just a few miles of canal to take me through Loughborough.

The day was damp and misty as I strode out between Birstall, on my left, and Thurmaston, both of which have given their names to locks. I was still in Leicestershire, still within reach of the city of Leicester, although the river was lost in the meadows, unaware of the life going on around it.

The old church at Birstall discovered a Saxon window last century but Thurmaston made an even more interesting and historic find when a workman dug up a Roman milestone which is now in the museum at Leicester. Fosse Way, the ancient Roman road, goes through Thurmaston, and Emperor Hadrian came this way on his visit to Britain in AD 120. The Roman milestone was dedicated to him and, as well as mentioning his father and grandfather, it pointed out that Ratae, known today as Leicester, was two miles away. It was unearthed in the eighteenth century and was the base of a lamp-post for nearly fifty years before being taken off to the museum. There are gravel workings with their accompanying lakes around here, all helping to encourage the wildlife and make the walking even more varied.

The road from Wanlip (with more Roman connections) to Syston crosses the Navigation at Bridge 19, where the Hope and Anchor Inn stands in all its glory, its tables, benches and moorings all new and unsullied. The River Soar has taken itself off at Thurmaston to go past Wanlip and its church, which stands close to the water. Here too is a Roman house in the fields by the Soar. Syston is on the other side of the canal, a little further away, the meeting-place for two railway lines, from Corby and Chesterfield as they head to Leicester. There is an interesting fifteenth-century church with an inscription from the eighteenth century in the churchyard which reads:

> What I have I gave,
> What I spent I had,
> What I left I lost by not giving it.

This area is a mass of water, with the Navigation, the river and the flooded gravelpits, and it is a nice change to have the canal again with the Old Junction Boat Yard at Syston and the narrow-boats there at rest. A footbridge here took me over the River Wreake, which comes in from the right and becomes the Navigation for the next mile. A small white milestone bearing nothing but the figure 18 – the number of miles to the end of the Grand Union at Trentlock – comes at Junction Lock, and at a sharp bend in the river is a four-foot-high pole, as thick as a telegraph pole, with a strip of iron round it to take the boat's tow rope. The Wreake provides delightful walking through the fields before a footbridge crosses the road of water as the River Soar prepares to take over the Navigation again at Cossington Lock.

Cossington is a peaceful, picturesque village about three-quarters of a mile east from the lock, the burial place of Lord Kitchener's father, who died in 1894 while his son was away in Egypt. On the other side of the water stands Rothley, with its temple, Anglo-Saxon cross and church, founded in Norman days. The temple takes its name from the Knights Templar who founded it as a preceptory in the reign of Henry III. The poet Thomas Macaulay was born here in 1800 but when he died, fifty-nine years later, it was Westminster Abbey which claimed his body, to lie alongside other poets. The house is now an hotel, a wonderful place in which to stay the night.

The meandering riverside path manages mostly to stay close to the water as it noses its way between Mountsorrel, which has the A6 running through it, and Sileby, an industrial village of little attraction. The magnificent tower of Mountsorrel's fourteenth-century church can be seen from the water just before the war memorial on the hill comes into sight. The small town is famous for its granite, and through the years many a barge-load must have left Jelly's Wharf for all parts of the kingdom. There were

Bridge over the Grand Union Canal near Old Junction
boat-yard at Syston

narrowboats there and under the trees an old barge lay at peace, perhaps after many years of carrying the pinkish product from which so many churches, houses and streets have been built. The path passes along the front of the Waterside Inn before continuing its winding way towards Loughborough. But first the Navigation has to pass Barrow upon Soar, one of several villages or towns to attach its name to this pleasant river in the same way as many have done alongside the Thames.

The river goes off on a great loop to Quorndon, while the

Navigation takes a much straighter course along the edge of Barrow upon Soar. Quorndon is a large village standing on the edge of Charnwood Forest and has been the home of the Farnham family for 700 years. There is, of course, a Farnham Chapel in the fourteenth-century church, with monuments to the family, notably a great altar tomb of John Farnham from Elizabethan times.

Barrow upon Soar is at its prettiest near the water, with houses reaching down to the bank, trees dipping their heads, boats and ducks all making an attractive waterfront. It was a lovely spot when L. T. C. Rolt came here in *Cressy*, so inviting, in fact, that he moored up and stayed for four days – four halcyon days, he called them.

There is no path across the weir where the river leaves for Quorndon, and it is necessary to follow the weir left and cross a barbed-wire fence and a wooden fence to reach a bridge not five minutes away from the Navigation. The path leads from the bridge to a lane at Meadow Farm, and a right turn took me back to the water and Shipstone's Navigation Inn. Barrow Deep Lock follows ten minutes later, with its traffic lights and notice: 'When green light shows it is safe for craft to proceed beyond this point.' The river takes over again at this point and enters an attractive wooded stretch as it makes its way to Pillings Flood Lock where the waters part company again, the River Soar taking itself off to the right, and the Navigation, soon to become the Loughborough Navigation part of the Grand Union Canal, going left. Now I was back with a real canal, straight and sure, knowing exactly where it was going.

Loughborough is not far away, only about two miles, but for the time being this is wide-open countryside with the more familiar hedges, fields and trees. But just when I was thinking I was miles from anywhere, a golfer appeared to the right, a lone golfer with a red-and-white umbrella as the drizzle threatened to take a hold of the day. The railway is to the right, and hills are to the left but only twenty minutes after leaving Pillings Lock the first houses and factories of Loughborough appear. It was another half-hour, however, before I really felt I had found Loughborough as I settled in at the Boat Inn on the canalside and looked out at the church of All Saints with its fifteenth-century tower.

There are one or two more notable buildings in Loughborough, the biggest town in the county next to Leicester, but my recollections from three nights spent there one summer were of a generally unattractive town. Yet Loughborough, it seems, is famed for its church bells, which have been made there since the middle of last century, and the town is responsible for some of the finest bells in the world, including one in St Paul's Cathedral. The back gardens of several houses run down to the water, and there stand family boats, at the private moorings, coming in all shapes and sizes and colours, like the family cars at the front of the house.

As the Navigation starts to leave town, it hits a T-junction with Loughborough Wharf and Basin to the left, the main line to the right, past Albion Inn, boarding kennels and a Do-it-Yourself supercentre, and on to Town Lock. Between the wharf and Town Lock is the official boundary between the Leicester and Loughborough Navigations, the Loughborough taking over for the remaining nine miles to the Trent.

By Bishop Meadow Lock, about fifteen minutes further on, the housing and industry have been left behind. Ahead are wires and pylons running with, and across, the water. The river joins up again just past Bishop Meadow Lock, bringing with it the boundary line between Leicestershire and Nottinghamshire. From here, for the rest of the day, Nottinghamshire will be on the right bank across the water, with Leicestershire on the left.

The spire on the right belongs to the Nottinghamshire village church of Normanton on Soar, still more than a mile away and clinging closely to the river. As the village nears, a new sign warns: 'Caution, Ferry Crossing 200 yards. Beware of chain. Stop if ferry in operation.' The ferry was not operating but I could see the chain leaving the water close to the church, and I spotted a funny little boat on the Nottinghamshire side. The church of St James, almost entirely thirteenth-century, is one of the county's two churches with a central tower, and its churchyard is almost on the river. It was known as the Boatmen's Church from the time when the river carried cargo barges, and its lectern is made from a tree trunk. Not that that was any use to me, seeing that the ferry was not running. Instead I had to make my way gingerly through a Leicestershire field containing bulls whose menacing stare was

St James's Church in the Nottinghamshire village of
Normanton on Soar

enough to make me jump over a barbed-wire fence and tear my trousers.

Soar Boating Club appears on the right before the river swings left, past boats and houses and the Rose and Crown, as it reaches out for Zouch, pronounced not as you might think but 'Zotch'. Here is another lock with meadows on either side, the good feel of a river with fields and trees and some pleasant wandering.

The river enters one of its prettiest sections here as it makes its way to the bump at Devil's Elbow near Sutton Bonnington, and on to Kegworth. Sutton Bonnington church spire shows up on the right, with the belching smoke of Ratcliffe Power Station beyond. The path is on the right-hand side of the river, the Nottinghamshire side, a stretch of river without a way across for about three miles from Zouch to Kegworth.

Sutton Bonnington stands near Devil's Elbow, a village – well, two villages rolled into one – sitting on the slope of the valley and looking across the river. Originally there were two separate villages, Sutton and Bonnington, and the church with the spire is that of St Michael's, Bonnington, dating from the thirteenth century and with a 600-year-old font. A less noticeable church, but still one of much historic interest, is St Anne's at Sutton, which does not have a tower but is from the fourteenth century. In the chancel is a fifteenth-century knight in his armour, an enormously tall man – seven feet if the figure is true to life. His feet rest on a lion, and he was called Old Lion Grey.

Not far down river from Devil's Elbow is the White House on the other side of the water and managing to sit on both the Navigation and the A6 trunk road, built no doubt by somebody with an eye for business. Unfortunately, the White House is no use to the walker as it lies over a mile from the nearest bridge at Kegworth, a small town lying across the border in Leicestershire. A church with a spire stands on the main road, and there is a story that when the spire was once repaired a steeplejack celebrated the completion of the job by sitting on the top and blowing his horn. And the villagers, no doubt, stared up in admiration. The Kegworth villagers must also have found a great deal to admire in one of their sons, Sir John Kirk, who had a hand, with Lord Shaftesbury, in founding the Ragged Schools which did so much work for poor children.

The end of the Grand Union Canal – or River Soar – is close at

The White House on the River Soar perched between the
Navigation and the A6 road

hand, particularly for the walker who can find no way across the water at Trentlock. In fact, in only a little over a mile, it is necessary to leave the riverside at Ratcliffe on Soar, close to the power station with its eight cooling towers. Here are trees and fields and pylons. The sound of the road is never far away, and aircraft heading for the East Midlands airport occasionally make themselves known. In the midst of it all, I saw a kingfisher.

I left the water at Ratcliffe, where the A453 from Nottingham to Ashby de la Zouch, crosses over. I headed for Ashby and after a few hundred yards turned right down a lane, and then right again as the M1 began to get uncomfortably close. That brought me out at Sawley Cut and a way across the water by the B6540, a road that runs to Long Eaton. The first turning to the right is the road to Trentlock, to the start of the second half of the journey.

The first half had been one of great delight, except for the section of the Grand Union with the overgrown towpath. The walking was easy, and the countryside was a joy, with hours and hours of peace and quiet.

OS MAP 129
Information Centres: Leicester (0533 556699), Loughborough (0509 230131), Long Eaton (0602 735426), Nottingham (0602 470661)

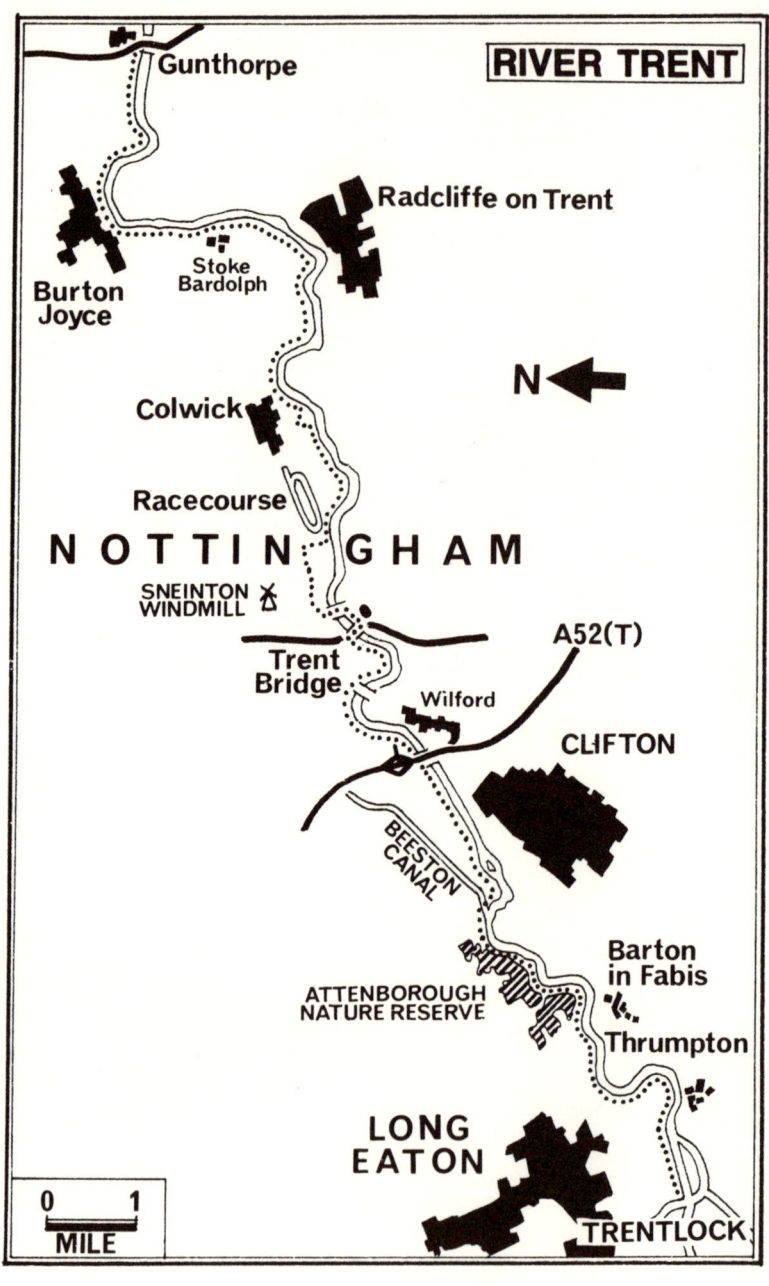

RIVER TRENT

Gunthorpe

Radcliffe on Trent

Burton
Joyce

Stoke
Bardolph

N

Colwick

Racecourse

NOTTINGHAM

SNEINTON
WINDMILL

A52(T)

Trent
Bridge

Wilford

CLIFTON

BEESTON
CANAL

Barton
in Fabis

ATTENBOROUGH
NATURE RESERVE

Thrumpton

LONG
EATON

0   1
MILE

TRENTLOCK

9

Trentlock to Gunthorpe

21 miles : 33.6 kilometres

The Trent is 180 miles long, the third of England's great rivers, behind only the Thames and the Severn. Trentlock marks somewhere around the halfway stage for a river born on Biddulph Moor in Staffordshire. It runs through Burton upon Trent, Nottingham, Newark and Gainsborough, before combining with the Ouse to form the Humber. It is a grand river on this stretch, a timeless, wandering, wide piece of water that keeps itself to itself for most of its journey.

At Trentlock the river hits a crossroads which is also the start of the Erewash Canal running north for twelve miles. The Erewash was promoted by Derbyshire and Nottinghamshire colliery-owners to transport coal from the Erewash Valley. The canal was completed in 1779, from Long Eaton to Langley Mill in D. H. Lawrence country. It cost £21,000 to build and was one of the most prosperous canals in the country, with trade from collieries, ironworks, brickworks and foundries. It has survived to provide boaters with a pleasant piece of navigable water, and walkers with a lovely day out.

It was a misty morning when I returned to Trentlock, a mile from Long Eaton and a shrine for the day-trippers who want to sit and stare at a busy junction that provides plenty of action at the weekends. The Trent Navigation Inn and the Steamboat Inn were at their quietest – not even the cleaners had arrived – and Ratcliffe cooling towers, only faintly visible across the water, were quietly filling the still sky with their smoke puffs. A huge sign pointed me to the River Trent and the North and as I set off I saw, across a field, a Union Jack fluttering faintly over a chalet, probably at the side of the Erewash Canal.

Soon after setting off I passed one of the very few locks on the Trent – leading into the short Cranfleet Cut – and went under the

railway line to Leicester, built by Stephenson. Somewhere behind me I could hear the rumble of traffic, lost in the mist that was
continuing to keep the cooling towers nothing more than a faint
blur. The sign behind me, motorway size, pointed to the Erewash
Canal on the right, the Trent and Mersey Canal straight ahead,
left, then right for the River Soar and the South. But ahead was
the river, a true river, and the Nottingham Yacht Club, with its
boats moored alongside, standing opposite Cranfleet Lock. The
river, below the lock, is about sixty yards wide, and its path is
through fields and well trod.

Trees line the opposite bank before the river runs through open
country on both sides to reach down to Thrumpton, where I came
across the water bailiff, who told me about the days when the
river quite regularly flooded the surrounding countryside. He
told me that in 1946 he bought a house in Barton Road and
decorated it throughout only for the river to flood and deposit

The motorway-type sign near Trentlock at Long Eaton

three feet of water in it. They cleaned up, but two months later the river went a bit higher, this time giving him 3½ feet of water. 'I thought at the time we were going to have this for ever,' he said, but six months later, when he was unable to use his bicycle because of a puncture and was waiting for the bus home from Kegworth, he was given a lift in a car. 'The driver turned out to be a Dutchman, in charge of a flood-control scheme that was being introduced, and when he heard where I lived he told me not to worry. I wouldn't have any more flooding. And sure enough, since then we haven't.'

The water bailiff told me there were not many boats on the river these days, and not many anglers either. He pointed out for me, through the trees, the lovely Thrumpton Hall. 'Byron lived there, you know,' he said. 'An old lady has it now.' There was a hall there from the eleventh century until the family paid for its part in the Gunpowder Plot in 1605, when they concealed Father Garnett, one of the conspirators, in a priest's hole. The house was virtually rebuilt in 1607, although parts of the original building were incorporated, and in more recent years two members of the family married Byrons, one of them Lucy Emerton Westcomb who lived to be eighty-eight and ruled the village.

The river performs a quite substantial loop here, and straight ahead, maybe a mile away, is the village of Barton in Fabis with its church spire. Barton in Fabis, once intriguingly called Barton in the Beans, is a charming village with thatched houses and farms and a grand old church, and was once an ancient ferry site. A small stone mile-post bears nothing but the figure '5½', which I took to be the number of miles from Derwent Mouth, just beyond Trentlock.

A quiet reminder that I had entered Attenborough Nature Reserve came when a heron lifted off the bank in front of me. A rather noisier sign came from about sixty Canada geese which had wandered down to the side of the water. The path continues on a piece of land about a hundred yards wide between the two waters, the river on the right and what is known as Clifton Pond on the left. Day tickets for fishing were advertised at 95 pence each.

I looked across the water where trees lined the bank and Barton in Fabis spire still thrust high. The Ratcliffe cooling towers were at last beginning to fade altogether, covered in their own smoke.

Terns were fishing in the area, darting and waiting, then diving
for the small fish they wanted. I came across a small sloping stone
wall which suggested there had once been an overflow but which
I now believe could have been the old exit of the River Erewash
into the Trent. Carved in stone was the request: 'Please close
gate.'

The river twists here and looks about a hundred yards wide
near where a sign points off to Nottingham Nature Trail, heading
in the general direction of the church at Attenborough, the
birthplace of John and Henry Ireton who served Oliver Cromwell
during the Civil War.

The path is good, though muddy at times, and the surroundings are beautiful. A man who was going from Beeston to Barton
Lane said he thought it was the best stretch of walking for ten
miles in either direction. Beeston Sailing Club headquarters are
across the river and, as I passed, a row of about sixty boats lined
the bank on the river bend up to the entrance to Beeston Canal
with its lock and lock-keeper's cottage. The river is so shallow in
parts here that it would be difficult for even a small boat to find a
way through, and that is why Beeston Canal was built.

Around here the factories of Boots the Chemist can be seen,
founded by Jesse Boot, who became Lord Trent. A statue of him
has been erected on the boulevard between Beeston and Lenton
with the inscription: 'Our great citizen, Jesse Boot, Lord Trent.
Before him lies a monument to his industry; behind him an
everlasting monument to his benevolence.' That benevolence
came in the form of the university and the splendid parkland with
its gardens, boating pool, paddling pools and sporting facilities.

The river heads off towards Clifton church tower on the hill
directly ahead, and the path crosses the canal bridge before
returning to the riverside. The hills and woods across the water
contain the Georgian mansion Clifton Hall, which stands alongside the church and is now a college. The river swings sharp left
here to run straight and fast along the foot of the cliff, tree-lined
and attractive. There was a ploughed field on the left, right next to
the path, and a warning sign to 'Keep to Footpath'. Not that there
was much path left, as the ploughing went right to the edge.

A feature of the Nottinghamshire part of the River Trent is the
white-painted clapper gates, double gates between the fields
with a leaning post. Part of the paths around here go through

muddy fields, and it was nice to leave them for the well-kept playing fields of the university for a little while. The river is still in one of its longest straight stretches as it continues to run along the foot of the hillside. To the left are cultivated fields with industry beyond, but it stays country quiet until the A453, across the water, starts to get near. The path moves out onto a road and passes Lenton United Cricket Club, Notts Gregory Hockey Club and Dunkirk Sports Club. I walked on, past Trent Side Farm, to a lane which leads under Clifton Bridge, a double bridge carrying traffic in and out of Nottingham. During the construction of this bridge in 1958 a great number of metal tools and weapons of the Bronze Age were found, in addition to three dug-out canoes from the river itself.

The suburb of Wilford is on the other side of the river, directly facing the power station with its church standing on the river's edge and presenting a pretty riverside picture that might look more at home back in the countryside. The city of Nottingham lies directly ahead but here the river turns its back on it as it starts an enormous loop. Deep steps line the bank on the city side of the river as it enters a park-like area with good, easy walking. Under the bridge facing downstream is the statue of 'Sir Robert Juckes Clifton Bt., M.P. 1861–69. Obit May 30, 1869. W. P. Smith Sculptor Nottingham'.

The loop which runs down to Trent Bridge is an impressive part of the river, a clean stretch with trees, lawns and steps leading down to the water's edge. Along here are the Meadows, twenty-three acres of land containing recreation grounds and open parkland with a splendid memorial gateway. It had originally been planned, after the First World War, to spend £20,000 on a cenotaph in the centre of the city, but so great were the protests that the Corporation agreed instead to provide the memorial gateway to the Meadows.

Here too can be seen three great sports grounds, belonging to Notts County and Nottingham Forest Football Clubs and Trent Bridge cricket ground, home of Nottinghamshire CCC and one of the world's most famous cricket grounds. The Notts County ground at Meadow Lane is on the city side of the river and is the home of the oldest League club in the world, having been formed in 1862 and becoming one of the original members of the Football League in 1888. Forest, on the other side of the river, stands close

Statue of Sir Robert Juckes Clifton, Member of Parliament from 1861–69, which stands on the banks of the River Trent at the edge of Nottingham

to the water and now holds a much more exalted place in soccer's élite than its neighbour. It was founded three years after County. Behind it stands the cricket ground, one of the senior citizens among grounds with its quite handsome pavilion, built in 1886, one of the finest in the country.

The ground was founded by William Clarke who, in the 1830s, married the landlady of the Trent Bridge Inn. Clarke was a bricklayer and a keen amateur cricketer – so keen, in fact, that he turned the site into the cricket field we know today. The inn was part of the club's ground for nearly a hundred years and was even used as the pavilion. Its successor, built in 1938, occupies the same site, and even today there is a way through the hotel into the ground. Customers also watch from its roof.

The open embankment, the delightful promenade, comes to an end at Trent Bridge, where there has been a succession of bridges for over a thousand years. It was an important crossing-point for centuries, the only one across the Trent between there and the coast, but it seems to have been regularly under repair or being replaced until the erection of the present one in 1871, designed by the Borough Surveyor, Marriott Ogle Tarbotton. It was strengthened and widened in 1925, and today seems to carry enough traffic to justify even further extensions.

Football and cricket are not the only sports represented near Trent Bridge, where, as one might expect, there are several rowing clubs, including NBRC, 1869, which overlooks the river close to one of the entrances to the Forest ground which proclaims: 'Trent End for Forest supporters only.' A firm of racing-boat builders and Nottingham Kayak Club are a little further down the road alongside the river, close to the next bridge, a three-span one called Lady Bay Bridge. This is the end of riverside walking for a couple of miles as industry stretches right down to the river's edge.

Nottingham is one of my favourite cities, with a warmth all its own. It has no pretensions. It is a working city which reaches back into history and which has played an important part in the development of modern Britain. Today's castle is relatively modern but a fortress has stood on the same spot for more than 900 years. A bronze statue of Robin Hood stands outside the castle, a monument to the hero of Sherwood Forest, first and most spectacularly displayed on screen by Errol Flynn. Nottingham is also renowned for its lace, John Player cigarettes, Raleigh bicycles and, of course, Boots the Chemist.

As I walked over Lady Bay Bridge to take the road to Colwick, I spied on a nearby hill a windmill. Its sails were intact and the building was so prominent on the skyline that I could not understand how I had never seen it on my previous visits to Nottingham. I discovered that the mill has only recently been restored after being derelict for more than thirty years. George Green built it in 1807 for grinding flour on Belvoir Hill at Sneinton, a village then with open fields, farms and several other windmills, but after working for over sixty years it fell into disuse. It was later used as a furniture-polish factory and caught fire in 1947, leaving it a burnt-out shell until 1979, when the George Green Memorial Fund bought the mill and presented it to the city of Nottingham. In 1986 it had been restored, and for the first time in over 120 years flour was ground in Green's Mill.

After going round and through an industrial estate I was fooled by a turning right off the Colwick Road named Trent Lane which led me to the Yacht Club but not to a riverside path. I turned back and kept going towards Colwick, turning right at the racecourse roundabout and heading for Colwick Park and the river, with the racecourse on the left. A plaque on a stone announced: 'Colwick Park. This park was opened by Her Royal Highness The Princess Margaret 20 November 1979.' The ruins of St John the Baptist Church contrast strongly with the park, with its lakes and marina, walks and horse-rides, with the River Trent alongside.

This is lovely walking through trim parkland. The city is well to the rear, and here is a fine example of the way a river can be incorporated into man's design in the country. No matter which way you enter Nottingham by the River Trent, you will experience just what man can do when he puts his mind to it – with a little help from Nature, of course. For at this end is Colwick Park, and upstream, on the other side of the city, Attenborough Nature Reserve.

Across the water rises the spire of the church at Holme Pierrepoint, a tiny village with just its church, an early seventeenth-century hall and some cottages. There is a way across the river at the Holme sluices, but not for the likes of you and me, I am afraid, just for those who are authorized. The riverside path runs out here, and once again it is necessary to leave the river and follow

A footbridge across the River Trent near the Meadows at Nottingham

the path back to the road. I walked past the Mobil depot and turned down Colwick Estate Private Road No. 5, along by the Esso depot and back to the river after about twenty minutes away.

Within three-quarters of a mile industry was behind me. Scrubland took over for a little while. A notice warned of overhead lines with forty feet clearance only. So I ducked. Another notice, much more weather-beaten, indicates the railway line, which comes teetering on legs on a bend to a three-arched brick and stone bridge. An iron span proclaims its makers: 'Clayton, Shuttleworth and Co., Lincoln 1850.' The lines once joined here to cross the river, and a splendid bridge with twenty-eight small arches stands impressively on the other side of the river from Radcliffe on Trent.

Radcliffe, so we are told, is named after the red cliffs which stand a hundred feet above the river and which provide extensive views and fine walks by the water. The present church is not much over a hundred years old but an inscription from the old church tells us of Stephen de Radcliffe, of the thirteenth century, who is believed to have been its founder. There used to be a figure of Stephen, carved out of oak, in one of the church recesses, but when the villagers heard about one of the Duke of Wellington's victories over the French, they dragged it out, dressed it up to represent Napoleon Bonaparte and set fire to it.

The river has to make a sharp left turn at the cliffs, straight up to the weir and the 150-yard run into the lock with its traffic lights and notice: 'When amber light shows, lock to be operated by the boater.' It was red. Another notice imposed speed limits – 6 m.p.h. upstream, 8 m.p.h. downstream and a mere 4 m.p.h. on the Sawley, Cranfleet, Beeston and Newark Navigations. I was heading downstream, so I put a bit of a spurt on, through the tidy lock area along the well-kept path out in the country.

It was hilly across the river, wooded and quiet on the towpath side, as I approached the village of Stoke Bardolph. The end of the village clings to the river, a cluster of colour-washed semi-detached houses, with a farm and pub separated from the Trent only by the road and a narrow grassy stretch. The river here looks all of a hundred feet wide; it is quite still and the start of a long, reasonably straight stretch running down to Burton Joyce. A sailing club has erected a raised hut at the roadside near the Ferry Boat Inn, and the popularity of the place is shown in the parking restrictions at weekends and Bank Holidays.

A ferry existed until recent years, close to the inn and linking Stoke Bardolph with Shelford, about a mile from the river on the opposite side. It is a pleasant village, quite small, yet historic, having once had a twelfth-century priory, dissolved four centuries later. During the Civil War Shelford belonged to the Royalists and proved such a nuisance that the Parliamentarians sent 2,000 men from Nottingham to seize it. Some 140 Royalists were slaughtered and the manor was burned down. There is also a story about a Shelford tailor early last century who made some rather nice, and not over-priced, waistcoats. Unfortunately he was also the sexton, and it turned out that he had been taking advantage of his having access to the vaults, where he had been removing the rich apparel of the dead to make fancy waistcoats.

The road from Stoke Bardolph to Burton Joyce runs nearly all the way in company with the Trent, and as the road leaves the river the path continues alongside the water. Then, as the water heads for Burton Joyce, with its church of St Helen and its railway station, it swings abruptly away, forming a huge horseshoe. For a while it runs close to the railway line which heads north-east for Newark and Lincoln, and south and west for Crewe, Nottingham, Derby and Birmingham.

This had been a perfect piece of walking ever since leaving the industrial estate at Colwick, and it was to stay like this right up to Gunthorpe. The River Trent is unarguably beautiful for most of this stretch, a wide and lazy river that makes walking an absolute pleasure.

Gunthorpe provided the first road bridge since leaving Nottingham some 4½ hours earlier. Road bridges, as I was to discover, were to be few and far between as the river headed for the Ouse and the Humber and its outlet to the sea.

OS MAP 129
Information Centres: Long Eaton (0602 735426), Nottingham (0602 823823)

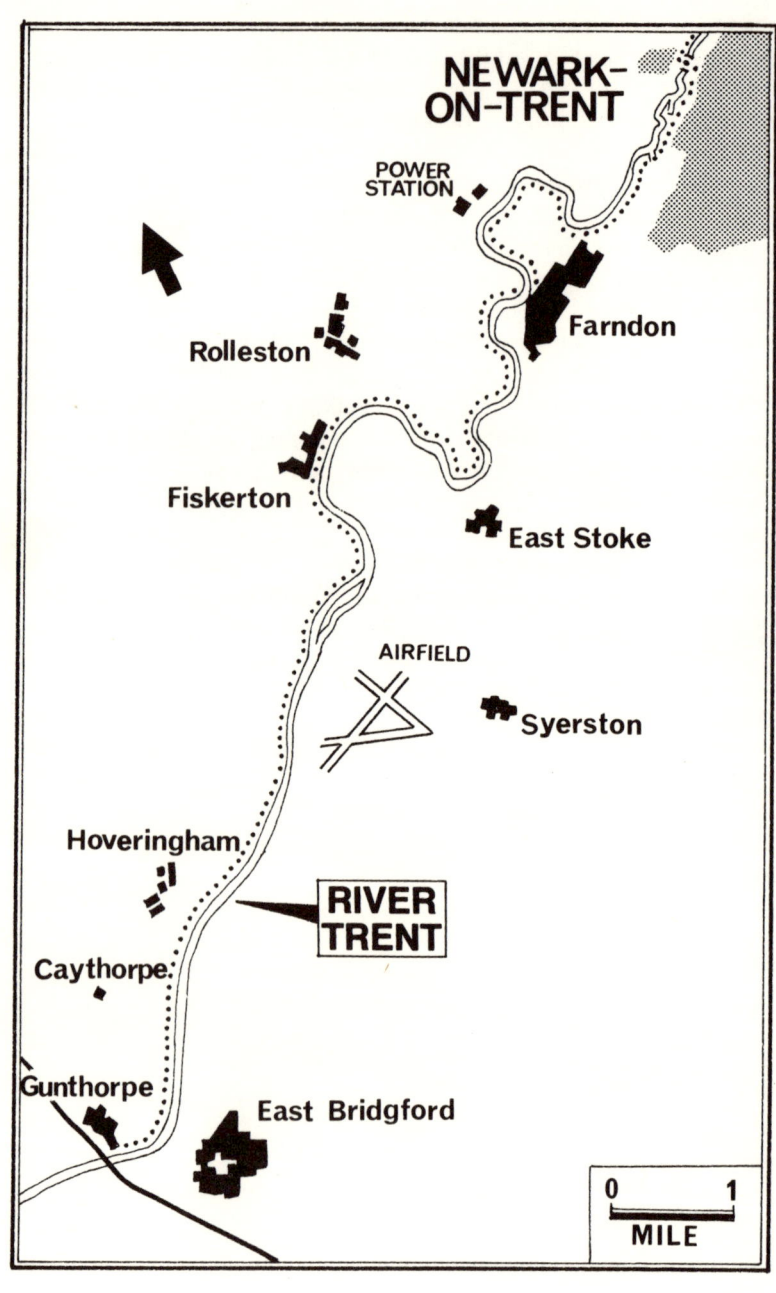

NEWARK-ON-TRENT

POWER STATION

Rolleston

Farndon

Fiskerton

East Stoke

AIRFIELD

Syerston

Hoveringham

RIVER TRENT

Caythorpe

Gunthorpe

East Bridgford

0   1
MILE

10

Gunthorpe to Newark

14 miles : 22.4 kilometres

There used to be a ferry at Gunthorpe, linking it with the village of East Bridgford. An iron-and-timber toll bridge took over in 1875, sixty-six years after the idea was first mooted to replace the boat. It remained a toll bridge for fifty years until the county council took it over, demolished it and built the present bridge, opened in 1927 by the then Prince of Wales, later to become Edward VIII and Duke of Windsor. The area by the bridge is a popular gathering-spot for picnickers, who can sit on the open riverside green and watch the comings and goings on the Trent. It is a pleasant area, even if the geese might be inclined to flap around, spitting, and there is refreshment galore, with two pubs, The Unicorn and the Anchor, and two restaurants, Tom Brown's and the Yeoman, all together within 200 yards.

East Bridgford is worth a visit, a fine old village which had a church in pre-Norman times and once had five public houses. The Old Hall was the home of the Hacker family, whose loyalties were divided in the Civil War. Colonel Francis Hacker was one of the signatories to the King's death warrant and subsequently lost his own head because of it. But his brothers, Thomas and Rowland, were on the other side, officers in the King's army.

Not far along the road from East Bridgford to Kneeton is the stump of a windmill which was built early in the nineteenth century and which lost its six sails when they were dismantled in 1939.

A weir with a lock marks more closely the site of the old ferry, and there used to be a wharf at this part of the river where coal and other goods would be unloaded. The river froths on its hurrying way down the weir as it swings gracefully away from East Bridgford and on past Glebe Farm, colour-washed in light khaki. This is a lovely reach of water with an easy path through the meadows by a lazy, lazy river, with hills and trees on the opposite side.

As I crossed Dover Beck I saw great numbers of what looked like Brent geese, about two or three hundred of them. Dover Beck runs for only a few miles, yet it used to feed five cornmills, a papermill and a cottonmill on its way into the Trent. The nearest was at Caythorpe, which can be reached in about ten minutes along the first road that appears on the left.

Caythorpe is a pretty village with attractive houses on the main street. The names fit in with the place: there is Ivy Cottage, Vine Cottage, Coach House and Manor House, with Bridge Cottage standing alongside Dover Beck. And there is Mill House, too, of course, a converted cornmill dating from 1749 with a stile and a footpath close by. St Aidan's is a small, corrugated-iron church – 'A chapel of ease of the Parish of Lowdham with Gunthorpe and Caythorpe' – not much bigger than a front parlour but with a brick-built extension going up. Opposite the entrance to Caythorpe Cricket Club, with its impressive scoreboard, is the bridlepath to Hoveringham, passing Hoveringham Mill Farm with its lovely red-brick buildings, one of which straddles the stream. It was perfectly quiet. No people, no sounds. Directions are there for riders to lead their horses between the buildings.

Hoveringham is another sleepy village with a church on a main street. Its quietness belies the fact that it is at the centre of the local sand- and gravel-quarrying industry. The mammoth is the trademark of Hoveringham Gravels because of the numerous bits and pieces of the extinct animal which have been discovered through the years. It is depicted on the side of the lorries, and there is also a full-scale model, made out of steel and standing twenty-three feet high, in position in front of the offices. Unfortunately the office and the mammoth were too far away from the village to encourage me to see them, so I made my way down Boat Lane and back to the banks of the Trent. Standing by the river is the Old Elm Tree Inn with tables in the garden overlooking the water towards Kneeton, on top of the cliff, with its quaint church of St Helen and the old tithe barn in Church Farm's yard. The Hoveringham ferry – one of dozens which used to cross the Trent – once ran from close to the Old Elm Tree Inn.

Sandpits and lagoons line the towpath side of the river, with a sign declaring 'Hoveringham Pastures' and prohibiting camping,

Standing on Dover Beck, a tributary of the River Trent, at
Caythorpe is Mill House, a converted corn mill built in 1749

cars and caravans. A footpath and bridlepath head for Hazelford
Ferry about two miles away, and Trent Power Boat and Ski Club
warn other water-users of the ski-zone. There were also notices
around here warning about deep water and quicksand, which did

not seem to worry several herons and partridges which flew out scared as I walked by. Now there was water on both sides of the path, the river gathering pace, the lakes as still as the grave. A glider floated serenely overhead, just like a heron, and the world was at peace.

The river was now truly majestic, approaching 150 yards wide, and the lakes had given way to flat, ploughed farmland extending for a mile. The herons could still be seen standing quietly on the river bank and although there were the occasional sounds of aircraft, the only signs that there was an airfield on the hill across the water were the soundless gliders. This was a bomber station during the last war – a commanding officer lost an arm while rescuing the crew from a Lancaster which had crashed here and burst into flames.

Hazelford Ferry is close at hand, with another fine-looking pub, a long, white building at the end of the road called the Star and Garter, an enormous place with a stable entrance. When the ferry was in operation this must have been a busy spot as it linked with Bleasby, less than a mile inland. Some of the ferry machinery is still there, and today the place is just as busy, if not busier, at weekends, when people flock to one of the most picturesque parts of the river for miles. Paulinus, Archbishop of York in the seventh century, is said to have baptized Christian converts near this spot. Across the river are the Trent Hills and Flintham Wood, climbing the hill and making the whole scene one of great beauty.

A few minutes further on are another lock and an impressive weir which has been designed to flow down two sides of a rectangular construction and carries the British Waterways Board warning sign: 'All persons traversing this weir in a canoe or other craft do so entirely at their own risk. The Board accept no liability for any damage to property or injury to persons arising from this practice.' It was all a huge undertaking, with concrete steps and traffic lights in the middle of nowhere. A footbridge links the towpath with the lockside, but again not for the walker, who still has some miles to go to the next bridge, at Newark.

The cooling towers of the power station at Staythorpe now come into view, and across the river, beyond the levelled-out land, is the village of East Stoke, which witnessed the bloodiest battle ever fought by the River Trent. It happened in 1487, two years after Bosworth, where Henry VII won the English crown with his defeat of Richard III. Now Yorkist conspirators, with Germans and Irishmen among their numbers, were heading for London and crossed the Trent by the Fiskerton ferry. Henry was waiting for them with a large army, and the rebel group was slaughtered by the river, more than 6,000 men dying in all, 2,000 of them Henry's supporters, in the three-hour battle. A stone memorial stands on private ground, bearing the words: 'Here stood the Burrand Bush planted on the spot where Henry VII placed his standard after the Battle of Stoke June 16, 1487.' The route to the ford, a narrow defile, became known as Red Gully because so many rebels were slaughtered there. Today it is called Red Gutter.

Less than 200 years after the battle the village saw large-scale death again, this time from the plague: 159 people died in a little over six months in 1646. The parish register records: 'There died in the town of Stoke, 1646, eight score and one, whereof of the plague seven score and nineteen.'

In the village of East Stoke and standing on a Roman road, the Fosse Way, is the Paunceforte Arms, named after a prominent family, including Julian Paunceforte who was the first British Ambassador to the United States. When he died in Washington in 1902, his body was returned to East Stoke for burial with the embassy flag draped over the coffin. The monument to him was erected in the churchyard but the Union Jack, the first that ever flew over a British Embassy in the United States, was hung in the church.

Across the river from East Stoke is the tidy, well-kept Fiskerton Wharf with the Bromley Arms. A flood-protection scheme has given the village a good defence against the Trent since 1959, and several handsome houses stand close to the bank, well walled with steps up to the garden, rather than a gate, in case of flooding. Fiskerton is an attractive riverside village of some antiquity with a mention as early as the year 958 and where, in 1270, the priory of Thurgarton was granted a market and fair. The Bromley Arms, Eagle House and Anchor Down are listed buildings dating from the eighteenth and nineteenth centuries. The boatbuilder's, which stands alongside, occupies an old malt-house, recalling an industry which was once important in Fiskerton.

The road and the river join up at the end of the row of houses and for a few hundred yards the road is the towpath. But as the road branches away left the path goes through one of the familiar leaning clapper gates and stays with the Trent in front of the house, the Old Wharf, Rolleston. It was here that I saw a clash between a pair of herons and crows – although it could hardly be called a clash, as the herons, among our most shy and wary birds, tried to get away. But their attackers, who could have been carrion crows, were persistent, harassing the herons and forcing them further and further upstream.

Just where the road and river part, sharp to the left, is Fiskerton Mill, set back from the road. The mill, a four-storey red-brick building, produces animal feed, and while the present building is largely from the middle of last century, following a fire in which five men lost their lives, it is probable there was a mill there in medieval times, belonging to Thurgarton Priory.

A seat on the river bank is dedicated to the memory of T. J. Jim Watson, 'a friend of the Nottingham Piscatorial Society 1985'. Another is in memory of Bert Elliott, honorary bailiff of the society, January 1971. It was rather cold for sitting, but I did. I sat at a part of the river where perhaps Jim and Bert enjoyed many hours and looked around at a familiar scene of open fields and trees and the river about seventy or eighty yards wide. But the power station hovers in the background, huge and dominant and attracting hordes of pylons, like pigeons flying home to roost. In one great semicircle I counted over fifty of them.

A few minutes downstream the river sweeps past the parkland of Stoke Hall, which can be seen through the trees 200 yards from the water with its colourful blue-and-white clocktower. About here the Fosse Way is at times less than a quarter of a mile from the river. Unfortunately it is on the other side of the water, along with the Roman settlement and fort at Ad Pontem.

The river twists its way towards the power station, and across the water the road comes close, bringing with it a huddle of anglers trying to keep out the chilling wind. The pylons cross the river here, aircraft whizz around on test flights and, as the river twirls its way through the countryside, the power station is forced to change position. One minute it is to the left, the next to the right – once or twice it almost managed to get behind me. The path is now on the banking, but it is still good, still easy walking, as it has been almost all the way from Nottingham.

The fourteenth-century church tower of St Peter showed up across the river, and it was around here that I noticed, not for the first time, that the path had changed banks without any apparent way across. It had happened around Holme Pierrepoint, where there had been no possible way across the river and where, for a while, I had had to walk on the 'wrong' side from the racecourse at Colwick. Why had the towpath to change sides? I inquired. Because at times a landowner would not give permission for the path to go through his land, and so it had to be switched to the opposite bank. And when that happened, the horse had to be ferried across the river by boat so it could continue its journey.

There was mention of a ferry at Farndon before reaching the power station, a ferry which operated regularly until recent times, mainly for the power-station workers who lived on the Farndon side. Without it they would have had to travel three miles using a footbridge or about seven miles by road. I had telephoned earlier to make sure the ferry was still operating, only to be told that, while there *was* a boat there, it was really only for the people who belonged to the sailing clubs on both sides of the river. I managed to beg a lift, otherwise I would have had to leave the river bank to find my way round Staythorpe Power Station to Newark, and that would have been quite a detour. As I approached the sailing club, a horse nudged me in the right direction towards the water's edge. The boat, a small one with an outboard motor, was coming across, carrying oxygen and acetylene cylinders. It was a hundred yards across and I willingly handed over the 50 pence charge, knowing I had been saved two or three miles walking round the power station.

The ferry took me across to Farndon where the Britannia Inn and the New Ferry Restaurant look out over the water. A few people had been drawn to the neat and inviting picnic area from where a footpath sign points to Newark. The Trent Sailing Club and the Newark Sea Scouts boat centre are nearby, and a footbridge took me across the entrance to Farndon harbour and its hive of boats, most of them resting up for the winter. The river here performs a real horseshoe bend, threatening to turn back on itself before striking off once more in the direction of Newark. Crop fields were close to the towpath which had become muddy but was still easily walkable.

On the other side of the river stands Staythorpe Power Station

The 800-year-old Newark Castle, one of several castles to be
seen on this walk

which had first been visible more than two hours earlier. Close to the end of the buildings are the concrete sluices, and the water which has been used in the cooling towers can be seen returning to the river, rather warmer than when it went in.

Just beyond the station is Averham Weir, which takes the main channel of the Trent off to the north of Newark, linking up again at the other side of town, near Winthorpe. The Newark branch heads off towards a disused windmill without sails before swinging left on its way into town. The mill is the only one in Nottinghamshire in which the bricks were used narrowside on, making the walls over a foot thick. A date-stone shows the mill was built in 1823 by James Dike, five storeys high with 198 brick courses.

The path goes through a field with horses, past the windmill and on towards Newark spire, passing in front of a row of houses named Trent Villas close to the water's edge. Anglers were out in force and, when I asked one of them the reason for the clapper gates with a leaning post, he suggested briefly that the workmen of the time were probably persistently drunk.

The path now leads away from the river and, as at Holme Pierrepoint and Farndon, pops up on the other side of the water. I made my way onto the road and crossed over the River Devon – pronounced Deevon, I understand – before forking left and then turning left over a bridge to return to the riverside. I walked up to the Town Lock, where the locks were under repair, and looked up at the huge lettering on the side of the buildings – Newark Egg Packers, Trent Navigation Company. But it all paled into insignificance alongside the majesty of the 800-year-old Newark Castle.

OS MAPS 129 and 120
Information Centres: Nottingham (0602 823823), Newark (0636 78962)

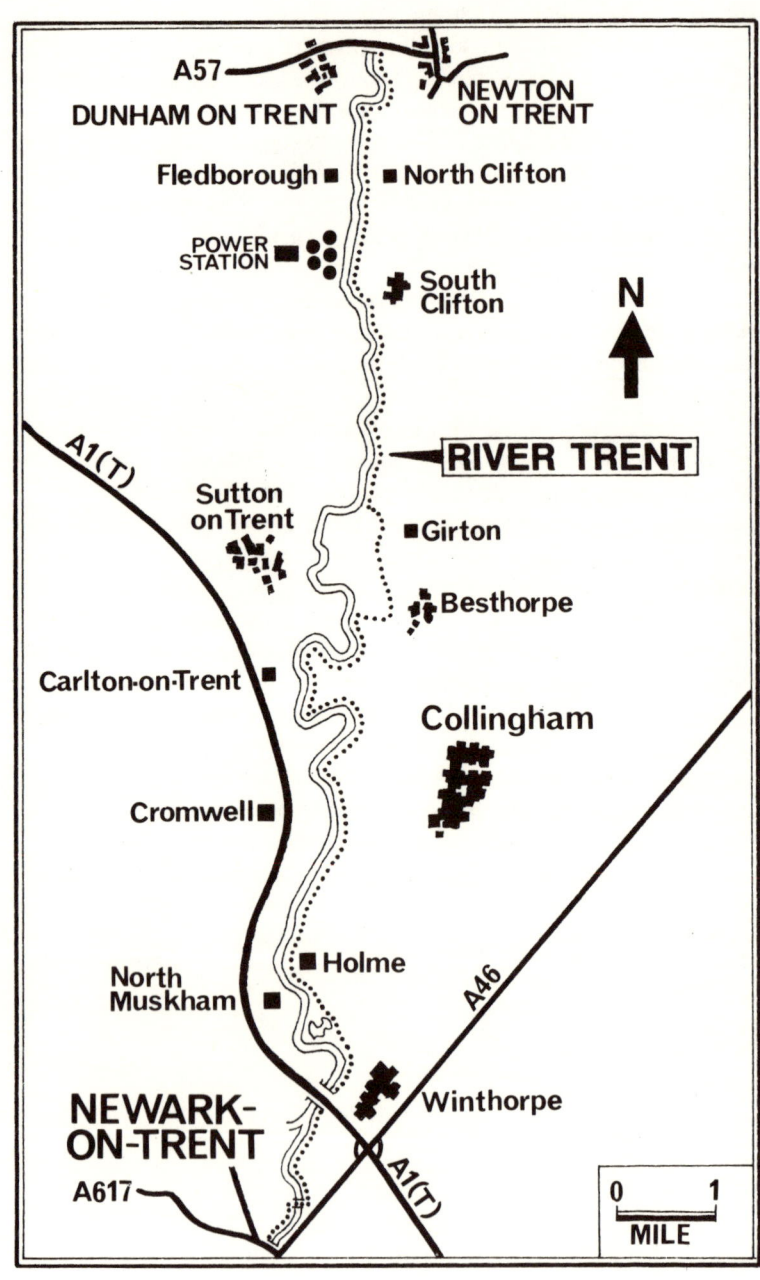

A57

DUNHAM ON TRENT

NEWTON
ON TRENT

Fledborough ■ ■ North Clifton

POWER
STATION

✚ South
Clifton

N

A1(T)

RIVER TRENT

Sutton
on Trent

■ Girton

Besthorpe

Carlton-on-Trent ■

Collingham

Cromwell ■

Holme

A46

North
Muskham

NEWARK-
ON-TRENT

Winthorpe

A1(T)

A617

0   1
MILE

11

Newark to Dunham on Trent

18 miles : 28.8 kilometres

Newark is a delightful place. It is small, with a population of about 25,000, yet it has a castle, one of the finest parish churches in England, and a cobbled market-place with a variety of old buildings. The Great North Road bypasses it today, but in the past it used to drive through the town centre, so that Newark was visited by, among others, most of the prominent diarists, even if they were only passing through. None of them had anything bad to say about the place, and most were quite complimentary. Daniel Defoe went there about 1720 and described it as 'a very handsome well-built town, the market place a noble square'. A few decades earlier, Celia Fiennes had called it 'a very neate stone built town. The market place is very large and look'd fine'. John Wesley said it was one of the most elegant towns in England, and William Cobbett thought it a very fine town.

The long wall of the castle looks out over the Trent and provides a handsome backcloth to the river. It is a real delight to arrive in the town either along the river or approaching it from the north and see this bit of castle that has been standing for 800 years. Unfortunately there is little else left of the castle, a disappointment that Defoe could not hide on his way from Leicester to Lincoln. He wrote:

The Foss Way leads us from hence through the eastern and north east part of the county, and particularly through the vale of Belvoir, to Newark in Nottinghamshire. In all this long tract we pass a rich and fertile country, fruitful fields, and the noble River Trent, for twenty miles togethere, often in our view; the towns of Mount Sorrel, Loughborough, Melton Mowbray, and Waltham in the Would, that is to say, on the Downs; all these are market towns, but of no great note.

At Newark one can hardly see without regret the ruins of that famous castle, which maintained itself through the whole Civil War in England, and keeping a strong garrison there for the king to the last, cut off the greatest pass into the north that is in the whole kingdom; nor was it ever taken, 'till the king, pressed by the calamity of his affairs, put himself into the hands of the Scots army, which lay before it, and then commanded the governor to deliver it up, after which it was demolished, that the great road might lie open and free; and it remains in rubbish to this day. Newark is a very handsome, well-built town, the market place a noble square, and the church is large and spacious, with a curious spire, which, were not Grantham so near, might pass for the finest and highest in all this part of England. The Trent divides itself here, and makes an island, and the bridges lead just to the foot of the castle wall; so that while this place was in the hands of any party, there was no travelling but by their leave.

The Bishop of Lincoln built the castle in the twelfth century to guard the crossing of the Trent, but it was not long before the Crown took it over. It was here that King John died in 1216 but, instead of being taken to the much nearer Lincoln Cathedral, his body was taken to Worcester for burial. Newark Castle was a Royalist stronghold in the Civil War and three times drove off the Parliamentary troops who tried to seize it. The garrison was allowed to march out with honour at the end of the war but the castle was razed – not, it seems, out of revenge on the part of Oliver Cromwell but because it was in such a state of ruin anyway. Happily, the face of the castle remains, with its arches and windows of Norman, Gothic and Tudor styles, and also the Norman gatehouse.

The market square, one of the largest cobbled squares in the country, is dominated, though not overwhelmed, by the town hall, built in 1773. The Queen's Head, dating from the late fifteenth or early sixteenth century, is to the right, and close to it is a pump built of cast iron and the original post where bears were fastened and baited. Beyond the National Westminster Bank, in Stodman Street, is the timber-framed Governor's House where the Governors of Newark lived during the Civil War sieges and where King Charles I quarrelled with his nephew, Prince Rupert, following the Parliamentarians' victory at Bristol. The importance of the house is also shown in the path which runs from it to the south door of the parish church, so the Governor could walk to church without getting his feet dirty.

The Clinton Arms Hotel was an important inn during the days of the post-coaches and had as many as ninety horses in its stables. Lord Byron stayed there when his first poems were published in Newark, and William Gladstone addressed the electors from its balcony when making his first election address in 1832, at the start of a career that was to see him become Prime Minister. The Saracen's Head Inn, where Sir Walter Scott stayed and where 'Jeannie Dean' stayed in his novel *Heart of Midlothian*, is alas no longer open, but the bust of a Saracen is still there, high in the wall. Another former inn is the Olde White Hart, a fourteenth-century timber-framed building which takes its name from the emblem of Richard II and which has tiny figures of angels and saints on its front.

The parish church of St Mary Magdalen is a little beauty. Nikolaus Pevsner described it as 'one of the two or three dozen grandest parish churches in England'. Arthur Mee reckoned it 'second to none' in Nottinghamshire. It towers benevolently over the market square, its early fourteenth-century spire going up and up for 242 feet, as high as the church is long. I wish I had had my binoculars with me because there is so much detail to see from outside. There are hundreds of gargoyles, figures and shields, and on one buttress are two men in a boat with another man pushing them off; two more men are quarrelling and one is pulling the other's hair. The Civil War destroyed most of the church's stained glass but what remains can be seen in the east window of the south aisle.

Another attractive building that takes the eye is the Ossington Coffee Palace, built in 1882 by Viscountess Ossington as a temperance hotel, across the road from the castle.

Here is the bridge across the Trent, standing where there has been some sort of crossing for hundreds of years. The first bridge arrived when the Bishop of Lincoln, who was lord of the manor, was granted a charter in 1135 '. . . that he may cause a bridge to be built over the water of Trent to his castle of Newark'. A succession of timber bridges, strengthened later by stone piers, gave way to one of brick, faced with stone, built by the Duke of

Parish church of St Mary Magdalen, Newark, with its
fourteenth-century spire overlooking a fine market square
which Daniel Defoe described as noble

Newcastle in 1775, which is still standing. The five-arched bridge which used to carry the traffic of the Great North Road is listed as an ancient monument, and the size of its arches, which cannot be altered, limits the width of boats that can use this water.

It is a shame to have to leave Newark, the finest town on the Trent. I can only recommend that anybody taking this walk – or perhaps just the River Trent section – should stay in Newark. It is worthwhile.

The way out of Newark, starting from the side of the river opposite the castle, is interesting, with the Castle Barge and some fine old buildings, a brewery and a maltings, warehouses and an old ironworks, all contributing to the lovely feel of the town. Soon after leaving Newark and before reaching the railway, it is necessary to cross the water through a double gate and over the white, arched footbridge. The path goes under the railway and past Newark Nether Lock before going under a second railway line, along a footbridge set into the stone pillars. An inter-city train rattled along as I reached the point where the rivers join together again after splitting at the other side of Newark four miles earlier.

The church tower at South Muskham can be seen about a mile away on the left, and the sandstone spire of Winthorpe's All Saints Church on the right as the path continues through a field towards the graceful bridge that now carries the Great North Road past, instead of through, Newark. The path crosses back to the left-hand side of the river at the bridge, the last road bridge until Dunham, nearly fifteen miles away. The Trent River Authority has established an island water fishery at Winthorpe, and the path passes by it, leaving it on the left and keeping to the banking as it heads for the church of St Giles at Holme, with its red-tiled roof and short, quaint spire.

There are no major towns alongside the river in its thirty-two miles between Newark and Gainsborough, just a succession of villages or hamlets, several of them close to the water. Many come in pairs, facing one another across the water and once linked by ferries which have long since faded into history. One such village is Holme, a pleasant spot with farms and orchards which was a hamlet attached to the parish of North Muskham until the sixteenth century, when the Trent changed course during a great flood and separated the two places. Before the flood Holme and North Muskham were side by side; today they must be nearly five miles apart, linked by the Great North Road bridge at Winthorpe.

Early in the seventeenth century the greater part of Holme belonged to Sir Thomas Barton, a man of great property in Lancashire. An ancestor of his built '. . . a fair stone house in the village and being a merchant of the staple placed in the windows this posie: "I thank God, and ever shall, It is the sheep that paid for all".' Tradition has it that Dick Turpin, the highwayman, frequently found shelter in a small, low cottage in Holme, the last in the parish on the way to Newark, which has since been taken down. Early in the nineteenth century several of his possessions were in the cottage, including a richly embroidered pistol-holster. Turpin is thought to have stopped only for refreshment for his horse while on the way to York, but the man who lived in the house, it was said, was taken to court for having harboured him, and received a severe sentence.

The church door was locked when I arrived but a notice told of several keys scattered around the village, including one right next door. I was happy to go in, out of the rain, into a charming little church with roomy aisles, two chapels, old pews, prominent beams and interesting windows. The two centre lights include twelfth-century glass from Salisbury and sixteenth- and seventeenth-century glass from Beauvais. I could hear the rain pattering on the roof, the only sound to break the peace and quiet. I felt I could have been in the fifteenth century; I wondered how much had changed within those church walls in 500 years. Very little, it seemed. The world had stood still.

E. G. Wake, in his *History of Collingham and its Neighbourhood*, published in 1867, referred to the excellent carvings in the old stalls and said he thought the original beauty of the chantry chapel must have been considerable. There are raised tomb effigies of Ranulphus and Eleanor Barton and underneath the almost skeleton-like son Ralph who died in 1579. Around the tomb in Latin are the words: 'Pity me, pity me, young friends, because the hand of the Lord has touched me.'

A little sign points the way to the steps up to Nan Scott's Chamber, a small room which cannot have changed in 300 years. I went inside and stayed for a few minutes, recalling the story of

Nan Scott and looking out of the same window she did when, during the Plague in 1666, she took refuge in the chamber, taking with her food for several weeks. There she sat, watching the funerals of village people, many of them her friends, until her food ran out. When she left the church to seek fresh food, she found only one other person living. Full of fear, she returned to the room and stayed there until her death. The box in which she is said to have slept is still there, and early last century some of her clothes and furniture could still be seen.

In 1851 a man called John Smith left sufficient money for ten poor people in Holme, North Muskham and Bathley and to provide each of them with a black gown and a dinner once a year. Also a convenient, handsome room within North Muskham church was to be placed at their service, where they might sit and hear one sermon every three months.

About a mile below Holme are Cromwell Lock and the weir where the tidal section of the River Trent begins its fifty-two-mile run to the Humber. The huge semicircular weir looked all of 130 yards across, an impressive structure with a lock, so I under-

Castle Barge on the River Trent at Newark

stand, big enough to take three giant oil-tankers. Cromwell village, on the other side of the water, was the home of the Cromwell who was Lord Treasurer and built Tattershall Castle during the reign of King Henry VI.

Public footpaths now run on both sides of the river, and across the water I could see the familiar white double-gated clapper gates marking the way. Over to the right is Collingham, once a divided village of South and North Collinghams but now, it seems, married into one with two churches. In 1795, when a thaw set in after seven weeks of frost, the whole Trent valley was left in desolation. According to Mr Wake, in his *History of Collingham*, 'Some persons now living [1867] remember a barge unloading coals into the churchyard after having navigated the lanes between the river and village at South Collingham.' That was the year of the highest recorded flood, and the mark of the height it reached is on the side of the base of a cross in the wall of the churchyard of All Saints, the parish church of North Collingham.

Evidence of a Roman bridge was uncovered during the dry summers of 1792 and 1793, when the Trent dried up so much that the foundations of a huge bridge became visible. Major improvements to the river in 1884 re-discovered two of the piers, constructed of masonry, which had to be destroyed to make more room for the waterway's commercial traffic. About a mile to the east of Collingham is Potter Hill, where between sixty and seventy skeletons, all with their thigh bones broken, were discovered in 1840. They had been buried three or four to a grave, some in a sitting position and some lying with their feet to the north-east.

Power stations seem for ever in sight now. The one at Staythorpe can still be seen to the rear, and also appearing are those downstream at Marnham, about eight miles away, and at Cottam, beyond Dunham on Trent, five to six hours' walking ahead. The country is getting flatter, windmills are popping up and, while flat countryside can be considered boring, it at least gives the walker a fine panoramic view. Some of the walking on the Grand Union Canal had been through cuttings or close to hedges, with restricted views, but here the countryside throws her arms open wide, like a mother welcoming home her child. There is so much to see, if only fields and farms, woods and village churches – and power stations. Some people might find power stations obtrusive, but they do not offend me. I like their egg-cup shape, and the steam hanging like balls of cotton wool in the still air, bringing a picturesqueness all its own to the scene.

The river here winds giddily and at one stage, though I had actually walked for three miles, I had gone no further forward than just one mile. The general direction was, of course, northwards but I might spend a quarter of an hour going east, another quarter of an hour going west, and at one silly period I was actually going south again, back towards Newark and Nottingham, not to mention Leighton Buzzard, Uxbridge and the Houses of Parliament. It was fairly frustrating. At least the footpath was straightforward – simply follow the river through the fields, past the cows. There is a windmill at Carlton on the other side of the river, one without sails, which seems to be in view for far too long because of the river's twistings.

Three anglers were up to their thighs in the river as I plodded on through fields without any obvious signs of a path. But when I at last passed the five-storey windmill, the path was better, and the gravel workings near the river at Girton could be seen before the river started to bend back on itself. The Redland works came into view, lorries rumbled past, and as I sat by the river to take notes a man hurried up in a car to see whether I was from the council, the ministry or some greedy rival organization. I found frequently through my walk that people regarded somebody with a notebook and pen with deep suspicion. Was it a guilty feeling, I wondered, like being reported at school?

The path does not go through the works and a few hundred yards from the gravel workings, it is necessary to go right up Trent Lane, a wide gravel road running between hedges. The first turning left leads to a gate with two footpath signs. I went straight on past two more, aiming for the cooling towers, walking between lakes, following a pulley taking sand. I had seen few dirtier, stickier places than this as I kept company with the pulley at the edge of a field until there was an opening and a gate on the right. I went through, then bore left along a grassy path leading to a track over a small bridge. In the hedge on the left, about a hundred yards before a 'road', is a footpath sign pointing to South Clifton, 2½ miles away. A series of stiles and yellow arrows led me back to the river and the road used by lorries.

Another gravel works – or was it the same one stretching on for ever? – stood in my way, but this time the path went right through it by the water's edge. Pulleys were carrying gravel to waiting barges at the wharf, and it was good to see signs of commercial traffic on the river. I had to duck under the pulley and felt the spots of gravel fall on my back as I pressed on to the bank to watch the barges being loaded.

The power station is now prominent, and low-flying test aircraft look to be lower than the cooling towers as they buzz about like bees. The power station itself is quite unmoved and quietly continues puffing like an old man with his pipe. The river performs another sharp bend, and the path wanders 150 yards from the water to cross a drain, the likes of which I was to see over and over again on the way to the Humber. The common sight on this river of two villages facing each other across the water, each with a lane running down to where a ferry used to operate, came again with the arrival of South Clifton, on my side, and High Marnham across the way. About a mile further on it was all repeated with the villages of North Clifton and Fledborough. Between them is the power station, linked to my side of the water by a suspended footbridge, alongside which is a well-occupied car-park for the workers from the east bank. There used to be one or two ferries around here, one of which gave the monks of Radford, on the order of King John, 'free passage for themselves, their servants and their carriage on his ferry boat'.

The people of South Clifton had the right of free passage on their ferry to High Marnham, although they did have to provide the ferryman with 'a prime loaf' each Christmas Day, when he and his dog by tradition had dinner at the vicarage. It was further understood that the parson's dog was always turned out while the ferryman's dog had his share of the entertainment. A rather nice custom, but one which has been out of use for many, many years.

Eleanor of Castile, wife of King Edward I, died at nearby Harby in 1290, and Edward, in his sorrow, ordered the erection of crosses at every point where the coffin rested on its way to London, some of which can still be seen. In another neighbouring village, that of Thorney, Thomas Otter was hanged in chains in 1806 for the murder of his wife on Christmas Day.

On the other side of the river is Fledborough with its delightful story about the rector who ran his own little Gretna Green there for about twenty years. His name was William Sweetapple and in the first half of the eighteenth century he granted licences and married young lovers on the same day without any questions. He conducted only six weddings in his first sixteen years at Fledborough but in the next twenty-five years, once he had got into the swing of things, he increased his score by 493! It has to be said in defence that he was made a surrogate in 1728, which gave him the power to grant marriage licences, a power which clearly went to his head and to the notice of young couples all over the land. A new Marriage Act in 1753 brought it all to an end, and with it, sad to relate, the death of the romantic rector.

As the river approaches the impressive Fledborough railway viaduct, with its fifty arches, it passes a church in isolation in the fields between South and North Clifton. A muddy line of path from the side of the church through the fields shows the way taken by the cattle to the edge of the river to drink. The footpath takes a curious turn at the viaduct, where a concrete footbridge goes out over the water and behind the pillars. The river flows gently here as it reaches Dunham on Trent whose toll bridge is still taking people's money after 155 years. It does seem archaic that one can still be charged to go across bridges on major thoroughfares.

OS MAP 121
Information Centres: Newark (0636 78962), Lincoln (0522 29828), Retford (0777 706741)

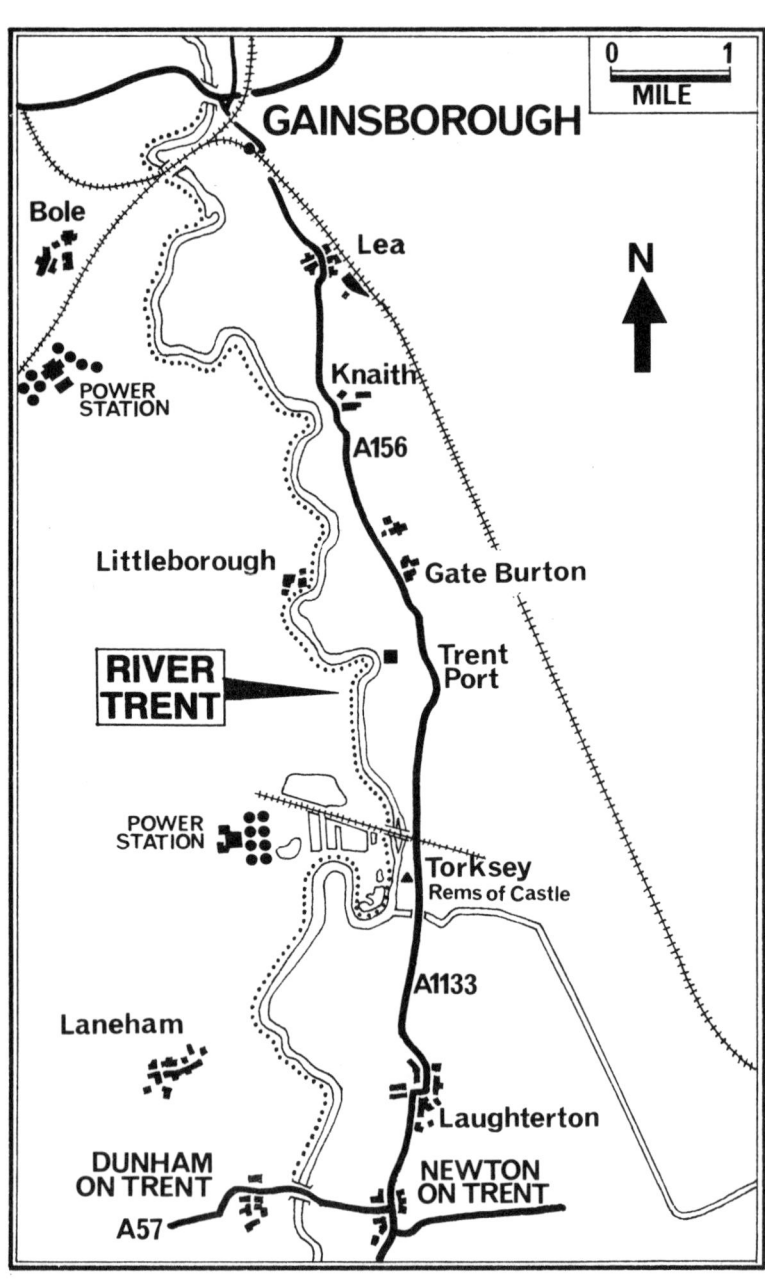

12

Dunham on Trent to Gainsborough

14 miles : 22.4 kilometres

On his journey to the Lake District from Cambridge in the summer of 1779, William Wilberforce, in his last years as an undergraduate, travelled through the Trent Valley:

> From Nottingham to Gainsbrough by the Trent, computed to be seventy or eighty miles. From Nottingham till one comes within four miles of Newark the Voyage answers very well, and there are a few pretty Views particularly at Colwick, Radcliffe etc., but afterwards there is little to attract a stranger's attention. The Trent about 10 miles from Gainsbrough makes two remarkable Bends which are called Burton Round and No Man's Friend. You are at the first of them within fifty yards of a Part of the River which you do not reach till you have gone near two miles. The Country about Gainsborough execrable, and the Inn we were at (which is said to be better than the other) miserably bad and dirty as indeed is the whole Town. I open'd the Window for a little fresh air, and the Smell which immediately filled the room was nasty beyond description.

From just before Dunham the river becomes the boundary between Nottinghamshire and Lincolnshire. Dunham itself is on the left bank, in Nottinghamshire, and is linked to the opposite bank by a toll bridge I have seen described as ugly but which has been scheduled as an ancient monument. Before the bridge was built there was a ferry at Dunham, and here in 1695 William III entered Nottinghamshire on a royal tour. But it was 'the dangers of passing a rapid river in the hazardous conveyance of a ferry boat', plus improved communications, that led to the construction of the toll bridge.

It is a long way to the next bridge again, fourteen miles to Gainsborough, and while the towing path is strictly on the Lincolnshire side for the first few miles, the footpath on the Nottinghamshire bank runs all the way to Gainsborough without a break and provides better walking. The path is on the banking a hundred yards from the river.

As I walked, far ahead the cooling towers of Cottam Power Station were busily puffing. The wind was only light but just about flattened the smoke and drove it across a sky which had the moon as high to the west as the sun was to the east. The willow trees on the riverbank on this stretch have been subject to the ancient practice of pollarding – branches cut back in an attempt to grow straight poles for building use. Some of the stumps can be hundreds of years old, but the pollarding has generally ceased now. The gently moving river wandered on, still far away from the path, which stays on the bank and provides excellent walking and views. The sights and sounds of the Lincoln road stay around for a while but the view is unending, with undisturbed fields for miles around.

Laneham church tower soon shows in the trees across the fields on the left, inviting the river to swing left towards it. The path has at last found its way to the waterside, still on the banking and heading towards Laneham village with its few red-brick houses, church, farms and the Ferry Boat Inn. The banking takes the path away from the river towards a farm but the route is to the right over a stile at the corner of the churchyard. The familiar yellow arrow is on the stile, leading into the graveyard where I spotted the headstone of Robert Newboult who died on 8 November 1835, aged sixty-nine:

> I was so long with pain opprest,
> That wore my strength away,
> It made me long for endless rest,
> Which never can decay.

The church of St Peter in Chains was locked – it is referred to in the Domesday Book as a 'Saxon church in a field just beyond the floodbank'. The present church, however, apart from the tower and the north aisle, was built early in the Norman period. I borrowed the key from the neighbouring farm, where bacon was sizzling in a frying pan, and returned to unlock a door which had been there for about 900 years.

The Norman doorway is beautiful, a fine example of the Early Decorated work of the time, and it was an experience just to stand there, in another part of England where time has stood still. Even rarer is the door itself, also Norman, which Arthur Mee described as 'one of the oldest in the land, one of the very few Norman doors we have opened and shut on our journeyings, perhaps a dozen in all England'.

Inside the lovely old church is another description of the door: 'As old as the doorway itself, its plain weather-worn boards are still good and it is swinging today on the same hinges made for it eight centuries ago.' The exact date is not known, it says, but 1110 would not be far out. There is a Norman font too, large enough for the immersion of quite a big child. The tower is 800 years old and was no doubt built as a watchtower on the Trent ferry, for churches were used in those days for many purposes besides worship.

The church contains an undated notice concerning Laneham parish which reads: 'The Poor have two shillings worth of bread every Sunday and a supply of coals in winter from the rent of two acres of Poors Land situate in the parish of Laneham and ten shillings yearly, in bread and money at Christmas, from the rent of a garden situate at the East End of the churchyard left by the Reverend Edmund Wallas to the Vicar of Laneham.'

I left the church through its fine old door and doorway and through the timbered porch that was erected this century. It was the work of local craftsmen, and hardly was it completed before the joiner's son died and had to be carried through the porch in his coffin. The wooden plaque in the porch states: 'In Grateful Memory of Ben Walker. One of the Builders of this Porch March 21st 1932. THE FIRST TO BE "CARRIED" THROUGH IT! July 2nd 1932.'

As I returned the church key to the farm, where the farmer was now well into his bacon and egg, I noticed road signs lying inside the church wall and obviously ready for use: 'Road Ahead Closed. Flood. Try Your Brakes.'

I walked down the road and past the Ferry Boat Inn, ignoring the Public Footpath sign to the river and following instead the road past the caravan site. As the road swings away from the

The church of St Peter in Chains at Laneham with its Norman
doorway and door which still swings on its 900-year-old hinges

river, a path follows close to the water's edge. I do not think I had been so close to the Trent since joining it at Long Eaton, and it is clear that this section must be flooded often. I was right next to the river, almost at a level with the water, and though parts of the path are covered, it is nothing serious and easy to bypass.

The path is soon back on a banking three or four feet above the river, and it is better at this stage to stay with the higher ground away from the flooded areas. The remains of a tree gallantly stick it out in the middle of the river, providing perches for cormorants scanning the water for food. Cottam Power Station is situated close to the part of the river where the water twists itself into a double U.

A barge called *Bustardthorpe* was loading up with sand at the power station wharf just as a British Waterways Board boat appeared, tugging a barge upstream. As I passed the wharf, I could see the church at Torksey not far ahead, but the river had first to turn back on itself, heading in exactly the opposite direction – due south – for about 600 yards before resuming its northerly course. As I headed south, now on perfect walking ground, I could see across the fields – and the river – the wall of Torksey Castle. Now I could see three power stations, all belching away and pouring their smoke into the sky.

A lake containing swans, geese and ducks is on the left as the river starts to turn itself round again, and there are only about thirty yards between the two waters. Just as the Trent swings itself back in the right direction, there, on the opposite side of the water, is the entrance to the Fossdyke and Witham Navigations.

The Fossdyke and Witham is one of England's oldest pieces of navigable water, forty-three miles from Torksey through to Boston, close to the Wash. For almost half its length it is in sight of the overpowering presence of the beautiful Lincoln Cathedral. The Navigations – for they are effectively two – provide miles of solitude, miles of walking or boating, in which you can travel without seeing a soul. The Romans were responsible for this stretch of water, having widened and straightened the River Witham and cut out the connecting Fossdyke. It is eleven miles from Torksey to Lincoln along the Fossdyke, which runs into Brayford Pool in Lincoln. The Witham then takes over, winding its way gently past villages and through the fenlands, past the banked-up entrance to the old, disused canal that ran to Horncastle ten miles away, and within a mile of Tattershall Castle. The Navigation ends at Boston, which has been a seaport for more than 800 years and which gave its name to the capital of Massachusetts after a band of Puritans left the town for America early in the seventeenth century.

It was the Fossdyke and Witham Navigations that helped the village of Torksey become so important in the Middle Ages, important enough to be established in the thirteenth century as the third town in Lincolnshire, behind Lincoln and Stamford. Ships carrying such commodities as corn, lead, iron, wool, wine, timber and fish plied to and from Torksey. At the time of the Domesday Book, 1086, Torksey had more inhabitants than Nottingham but by the fifteenth century decay had set in.

The great manor house of Torksey was the home of the Jermyn family in Elizabethan times. It was seized by a group of Royalists from Newark in 1645 who, after capturing 140 Parliamentarians, then set fire to the building. It was never rebuilt, and today all that stands, as at Newark, is the defiant face of the building, nothing but its front, with white stone at the bottom and red brick above. Its windows look out from the edge of the river across to Nottinghamshire. Once there were three churches in Torksey but today just one remains, close to the ruined castle.

The Trent passes beneath the railway bridge, which still has cross girders intact and, though clearly dangerous, is probably still used by people wishing to cross the water. The line on either side of the river is still used, with coal trains running to Cottam Power Station and oil trains to the oil terminal which stands alongside the river. The river continues to wander its way northwards, providing a good path on the banking, which is still making its way through white, leaning clapper gates.

Now the power station at Burton was taking up the skyline as I approached the rather grand-sounding Trent Port, which proved to be nothing more than a wharf for the nearby small village of Marton, with its church with a Saxon tower and windmill without sails. A British Waterways Board boat, *Robin Hood*, stood quietly at the jetty. Peace was all around. I was startled even to see an angler, for I had seen hardly any for a day or two. This is an area of ploughed fields, of flat – I suppose typically Lincolnshire – land, visible for miles to the walker if not to the boater, who would find it all hidden by the floodbanks.

Less than two miles downstream from Torksey is the hamlet of Littleborough, now quiet and cut off at the end of a lane which finished at the river but which knew busy and exciting days hundreds of years ago. Here was where the Roman road from Lincoln to Doncaster and York crossed the river, and here King Harold crossed with his troops as he went to his doom at Hastings. There was a ford here, constructed by the Romans in the time of Emperor Hadrian, who laid stones secured with oak piles. In 1820 it was ripped up because it was obstructing shipping, but during a drought in 1933 the remaining signs of the paved causeway could be seen. The road was Tillbridge Lane and the Roman settlement, perhaps a resting-place for the soldiers, was called Segelocum. As I stood at the water's edge, alongside Ferry Farm, I found it hard to imagine it being forded at that point.

The small church at Littleborough was erected by the Normans and named after St Nicholas. There are Roman tiles built into the wall, and other Roman remains that have been found here include altars and urns, pottery and coins. But the most spectacular find came in 1860, when the sexton was digging a grave and struck a stone coffin. The lid was raised and there was the perfect body of a woman whose burial robe was fastened by a Roman brooch. But she was not to be seen for long, for almost as soon as she was released to the air, her body turned to dust.

Over on the hill to the right, in Lincolnshire, is the village of Gate Burton with its Georgian hall. The parkland of the hall stretches down to the river, and on a bend, standing prominently on Red Hill, is a summerhouse, looking small and out of place. It was built at the same time as the hall, about 1765, and faced towards Littleborough. Behind it are the cooling towers and another interesting building, also in Lincolnshire, called Knaith Hall. It is an Elizabethan building, originally of red brick but now with a black-and-white front, inexplicably added in Victorian times. Like the gazebo, which is known locally as Burton Château, Knaith Hall is on a bend of the river, impressive and eye-catching with its tall chimneys – and having a Lego look about it.

The Trent now enters a fairly long, straight stretch of over half a mile before swinging back towards the power station with the village of Sturton le Steeple to the left and Lea to the right, across the river. It was this area that William Wilberforce referred to in 1779 when he wrote of the remarkable bend in the river called Burton Round. Yet only thirteen years after Wilberforce's visit *The Leicester Journal*, reporting a flood that had taken place in the same area, said: 'A very singular event has lately taken place at Gainsborough in Lincolnshire. At Bole Ferry the Trent has formed itself a new channel through which on Thursday se'nnight, two vessels passed abreast. Eighty or ninety acres of fine pasture land, the property of Sir E. Anderson and Miss Hickman, are cut quite away from the Lincolnshire side of the river and a complete island has formed between the late and present channels.'

Burton Round is still shown on the Ordnance Survey map, alongside the power station where the ancient church of West Burton used to stand. About a mile to the north-west is the village of Bole, once close to the river and with a ferry, but now about a mile from the Trent across Nottinghamshire fields that were once in Lincolnshire.

At last the river gets alongside the power station, which looks quite ugly at close quarters with surrounding land that is churned up and scarred. The fresh wind was flattening the smoke from the cooling towers and when I turned round I could still see the towers of Cottam Power Station four or five miles away. There is a road alongside the river from the power station but the path, up on the banking, is still good.

As I neared Gainsborough, a huge flock of starlings, thousands of them, rose from a field. They circled the area, glittering in the sun and looking like confetti that had just been thrown and was ready to settle.

The Spillers works stands right by the water on the opposite bank before the river runs under the railway bridge. The path goes with it under the bridge but a pair of stiles also lead across the railway lines and back to the river.

Near Gainsborough bridge, a man who was walking his dog by the river told me that Spillers used to get their deliveries by boat – big boats, coasters, but low-lying to get under the bridge. He told me the works next to it made breadcrumbs. 'You wouldn't believe the amount of breadcrumbs that goes out of here to all parts of the country,' he said. 'For fish fingers and all sorts of things.' He told me of the swans that used to gather nearby. 'We used to get two or three hundred swans here for donkey's years,' he said.

A modern, and common, sight on the River Trent, is the cooling towers of the power stations. This one is at Cottam

'Families used to come here with their kids to feed them. Then suddenly they vanished – I don't know where to. We used to get loads of barges up here and I remember one day a barge man telling me that seventy of his mates had been laid off in one week.'

OS MAP 121
Information Centres: Lincoln (0522 29828), Retford (0777 706741), Gainsborough Library (0427 4780)

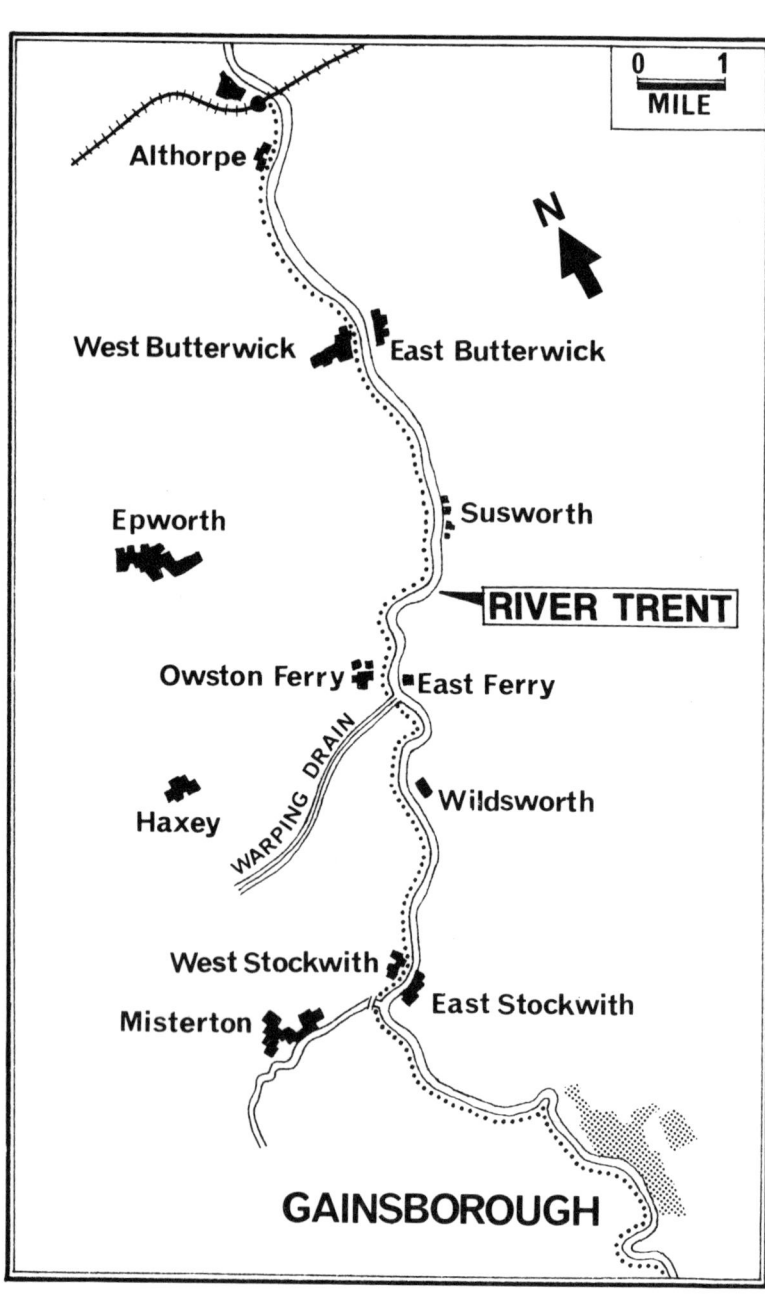

0   1
MILE

N

Althorpe

West Butterwick   East Butterwick

Epworth   Susworth

RIVER TRENT

Owston Ferry   East Ferry

WARPING DRAIN

Haxey   Wildsworth

West Stockwith

Misterton   East Stockwith

GAINSBOROUGH

13

Gainsborough to Althorpe

16 miles : 25.6 kilometres

Gainsborough has had a lot of uncomplimentary things said about it. It has been described as dull and dreary and even grim, with drab and gloomy thrown in for good measure. Its factories and warehouses have been termed forbidding, and some say it is best seen from the river. Its lovely old manor house, Old Hall, however, is exempt from all criticism, except perhaps for its surroundings.

Old Hall, from the fifteenth and sixteenth centuries, stands in the centre of the town near the library. The original mansion of Sir Thomas Burgh was destroyed in 1470 by the Lancastrians during the Wars of the Roses but he had it rebuilt in red brick in time to entertain Richard III there in 1484. Another of our more infamous kings, Henry VIII, also visited the manor house – in 1509 and 1540 and local tradition says he met Katherine Parr, who was eventually to become his sixth and last wife. The Hickman family lived there from 1597 until 1720, when they removed to Thonock, leaving their old house to go through a series of escapades as it turned into a chapel, a theatre and a soup kitchen among other things. It also served as a factory, a pub, a dance hall, an auction room and a Masonic temple before a series of renovations started on the hall in the late nineteenth century. There is an enormous fireplace at each end of the kitchen and great brick ovens which must have come in handy during the winter of 1816, when soup, biscuits, coal, herrings and potatoes were distributed to 435 families.

Gainsborough, like Torksey, was used as a port of entry by Danish invaders in the ninth, tenth and eleventh centuries, and it was in Gainsborough in 1014 that Sweyne of Denmark died after being recognized as King by northern and eastern England. His younger son, Canute, succeeded him, to add the crown of

England to those of Denmark and Norway, which made him the most powerful ruler in northern Europe.

Gainsborough's trade grew in the sixteenth and seventeenth centuries, and the eighteenth century saw the growth of the great brewing firms of Burton upon Trent exporting ale to Russia and the Baltic. Casks of ale were taken by local carriers by barge to Gainsborough where they were trans-shipped into brigs and carried down river to Hull. In the seventeenth and early eighteenth centuries Gainsborough's trade had been largely coastal but by the nineteenth century foreign trade figured more in the port.

The first steam boat seen on the Trent was the *Caledonia*, which made her maiden voyage from Hull in 1814 and caused great excitement in the district as it managed to move with a chimney for a mast. In 1815 Gainsborough shipyards began to build steamboats, and the first was suitably named *The Trent*, a wooden paddle-steamer for use between Thorne and Hull. It had taken three to seven days to get from Gainsborough to Hull by sailing packet; now the journey time was down to an almost unbelievable five hours. In 1822 *The British Queen* left Gainsborough for Hull at 9 a.m. and, after unloading cargo in Hull, was back in Gainsborough by 7 p.m., a round journey of 112 miles in under twelve hours, including stops at all the ferry places. By the 1830s there were regular steam-packet journeys between Gainsborough and London, taking a day and a half and costing 4s 9d.

The scene for George Eliot's book *The Mill on the Floss* was taken from a building which stood by the bridge but which no longer exists. She wrote of the quays where 'black ships laden with fresh-scented fir planks, with rounded sacks of oil-bearing seed, or with the dark glitter of coal' were to be seen. There are several proud, handsome warehouses here, too, some from the eighteenth century, and the river front is truly one of the best parts of Gainsborough. The bridge was erected in 1791, up to which time there had been no road crossing downstream of Newark, and it linked Gainsborough, which is wholly in Lincolnshire, with the Nottinghamshire side. It was declared free of tolls in March 1932 by the Minister of Transport, who paid the last toll on the bridge, using a coin dated 1770, as the first traveller did.

If I had timed the walk right, I could have witnessed the Trent oegre, the tide which rushes up the river in exactly the manner of

the Severn bore. The oegre's times are published in the news-
papers, and people gather to watch the phenomenon, which is
often at its best around Gainsborough. The oegre begins to make
its appearance very gradually below Keadby, about seventeen
miles downstream of Gainsborough, with what the sailors
termed a 'gentle shuft', and gathers strength as it rolls towards
Gainsborough Bridge. The Revd W. B. Stonehouse, in his book on
the Isle of Axholme published in 1839, said that when the oegre
was expected at Ferry – that is at Owston Ferry about halfway to
Gainsborough – boats pushed from shore into deep water, and
craft were all manned, with the steersman standing at the helm
waiting for the 'white curling wave accompanied by its rushing
sound'. Up would go the familiar cry of 'Ware oegre!' and
everybody was ready to lend a hand should the vessel drag her
anchor or, if she were heavy laden, get swamped by the swell.
'Vessels then got ready to sail back with the deep water to the
Humber,' wrote Stonehouse. 'If the wind was favourable, as the
tide continued to flow, brigs, schooners, sea sloops and keels
passed in rapid succession, so that on a fine summer's morning or
evening, at which time from six to nine o'clock the spring tides in
this part of the river always flow, the sight is truly animating and
delightful.'

In those days, the Trent abounded with freshwater fish, such
as roach, dace, bream, pike and most excellent eels. There were
also salmon fisheries at West Ferry, Kelfield and Gunthorpe, and
when fish was plentiful salmon sold at a penny a pound.

I left Gainsborough and crossed the bridge, knowing that again I
had all of about sixteen miles walking before I could get back to
the other side. I had returned to the Nottinghamshire side of the
river, but on the other bank I could see the church tower and the
Guild Hall. The tops of the chimneys of the Old Hall were
stretching up and peeping across the water. The wharf buildings are a good variety of shapes and sizes, rich in history
and providing the waterfront with character. There was some
character on my side, too, as I left behind the Trent Port Hotel,
standing close to the river and the bridge, and saw in a field a hut
carrying the sign 'Goat House'. There were plenty of coasters and
cranes and activity as I walked on, through a yard where timber
was being unloaded.

Gainsborough Old Hall, the manor house which dates from the
fifteenth and sixteenth centuries

The path soon takes to the banking again although the sight of
the river fades behind shrubland for a while. Morton village, with
its warehouses and church tower, appears on a little bump in the
river before the path moves fifty yards away from the river to
cross a stream by a Water Authority building. Walkeringham
village lies a mile away to the left, Misterton ahead and Walkerith
across the river, with its low farm buildings and houses with
greenhouses. The path on the banking is excellent, making
walking a pleasure, although the pylons close in for company
again. A large riverside works with a wharf, heavily protected
with barbed wire yet showing no signs of life, heralds the village
of West Stockwith.

Goats and hens watched me enter the village which marks the
start of the Chesterfield Canal and the end of the River Idle. West
Stockwith, like nearly all the villages on the Trent, is remote,
particularly now the ferry no longer plies to and from its sister
across the water, East Stockwith. The two villages are close
enough to hold a conversation across the Trent but, with the

nearest bridge being at Gainsborough, they are actually ten miles apart by road. The Crown Inn has a welcoming look about it, and West Stockwith Yacht Club has a good viewing position at the basin, alive with all sorts of boats.

The Chesterfield Canal was among the early ones, opening in 1777 and running forty-six miles from Chesterfield through Worksop and East Retford to West Stockwith, where goods were transferred to larger vessels. It was built to provide a better outlet through Nottinghamshire to the Humber for the products of North Derbyshire and South Yorkshire. Before that, the goods went by pack mule to Bawtry, then by boat. Coal became the canal's main carrying commodity but other cargoes included stone, corn, lime, lead, timber, iron, pottery and ale. The last commercial traffic was from the brickyards at Walkeringham to West Stockwith in 1955 after 178 years of cargo-carrying.

I pottered round West Stockwith harbour for quarter of an hour before setting off to go through the village, which is stoutly protected by walls and banks from the threat of the Trent. I called in at the church of St Mary the Virgin to see the monument to a ship's carpenter. He reclines close to the wall: 'Here lies the body of William Huntington, Late of this Place, Ship Carpenter who by his last will and testament after ye death of his mother and the marriage or death of his widow gave £740 for ye building of ye chappel and hospital round about it. And for ye support of a minister, schoolmaster and ten poor ship carpenters widows and other charitys, bequeathed all his lands in West Stockwith, Gunhouse and Misterton for ever. He dyed December the 24, 1714, aged 41.'

Close by is the end of the River Idle, after running forty-six miles, and Mother Drain, the two of them having worked their way to the Trent together for many miles.

As I was leaving the village, a man told me he had worked for thirty-one years at the locks on the river.

'I should know what I'm talking about,' he declared. 'This river shouldn't flood, but I bet it will one day. There are weaknesses, like here and at Gainsborough. I came here in 1946 and there was a flood that winter. I escaped, just, but I worked five weeks, day and night, getting rid of the water. It hasn't flooded since and they say it won't, ever again. But I'm not so sure. The river has behaved itself since then; it's been very good. I used to see lots of boats aground, but not now. Still, I wouldn't be surprised if it did flood again.'

I walked on, past potato fields and a boarding kennels, on the road by the river, but unable to see the water because of the banking. I soon regained the bank near Gunthorpe, where the wall is five feet high in places to keep out the river. From a seat on the bank I watched a barge, *Humber Monarch*, low in the water, heading downstream. I looked across the river and over the fields to Owlet Plantation and Laughton Common before going through Gunthorpe, a village stretched out along the road and the river. Roads run on both sides of the river now all the way to Althorpe, through country which was once exclusively Lincoln-shire until the borders were shifted.

The Trent used to mark the boundary of Nottinghamshire and Lincolnshire up to West Stockwith. Then Lincolnshire claimed the River Trent, a few miles of its west bank and the whole of the east side right up to the Humber. For a thousand years, until 1974, Lincolnshire measured something like seventy-five miles by forty-five and was England's second largest county (the largest being Yorkshire), with the Humber its northern limit. Today, since the boundary changes, Lincolnshire stays firmly on the east bank of the river and has been so decimated that it cannot include even Scunthorpe and Grimsby among its towns. Nottingham-shire, South Yorkshire and Humberside have taken over the west bank, which includes Epworth, the birthplace of one of Lincoln-shire's famous sons, John Wesley, and the surrounding district of the Isle of Axholme, a fascinating, historical, solitary, indepen-dent place. The island was one of the first landing places of the Danes, and in 797 a great fleet came into the Humber, plundered the whole country from there to the Trent and returned home triumphantly with the booty.

This is the real fen country, and the island was – I suppose still is – bounded by the Rivers Don, Trent and Idle. It was drained from the marshland and turned into fine agricultural land by a Dutchman, Cornelius Vermuyden, in the reign of Charles I. But the local people fought it, rioting, attacking the Dutch workmen and causing considerable damage to the drains. I have heard it suggested that Folly Drain and Paupers' Drain are so named because of the feelings of that period. Before the area was drained, villages and smaller towns stood on high ground along the spine of the island, whose inhabitants were known as Isolonians. In the early eighteenth century the main crop was flax because of the area's dampness, but today it is dry enough for wheat and carrots.

The name 'Isle of Axholme' means 'Island of Haxey', which is still one of the principal settlements and the scene of the Hood Game, held every 6 January, which recalls the day, hundreds of years ago, when Lady Mowbray lost her hood, and local boys fought over it. Today boys and men from Haxey and the neighbouring village of Westwoodside compete, rugby-scrum style, for the 'hood', a leather cylinder, and to get it to one of the village pubs.

Wildsworth, with its red-tiled house roofs, straggles along the opposite bank before Owston Ferry, one of the most interesting of the villages in the Isle of Axholme, comes into view. I was still on the bank, taking in the view, when I saw the windmill of Owston Ferry, between the nearby road and the river and without its sails. Windmills first appeared in Britain in the late twelfth century and soon became popular in Lincolnshire. Once there were as many as 700 working in the county, most grinding flour, many for drainage until steam engines arrived. Now only a handful still operate and the area seems to be littered with derelict ones.

Owston Ferry was the port of the Isle of Axholme, and in the last century large quantities of produce such as corn, potatoes, carrots and onions were sent off to the London, Cornish, Yorkshire and Lancashire markets. Kinnaird's Castle used to stand here and, though it was not mentioned in the Domesday Book, it was dilapidated in 1174, having been destroyed after being taken by King Henry II. On the ground which once formed the castle yard now stands the parish church of St Martin, dating from the thirteenth century and built partly from the castle ruins. Low Melwood, which stands about a mile away by the road to Epworth, was the site of a Carthusian monastery founded in 1396. On the same site there was once an ancient chapel known as the Priory in the Wood. The village had a market and fair from a charter granted by Edward III, and stock markets, commonly called Ferry Fairs, were held there, with 'A Hiring for Servants' on 24 November each year. The markets and fairs stopped being held between 1868 and 1876.

The harbour at West Stockwith, a few miles downstream from Gainsborough

In the reign of Queen Elizabeth I it was agreed between the inhabitants of East Ferry, across the river, and the vicar of Owston that they should pay to the said vicar fourpence for their hens, and the inhabitants of Cottinger threepence, in consideration of which payment he was to christen their children, church their wives and administer to them the Holy Communion. The book containing the Owston churchwarden's accounts from 1660 to 1684 includes an item for the year 1663: 'Paid to Edward Terwitt's wyfe for pulling Everatt's boy's head, five shillings.'

Owston Ferry was originally called West Kinnard Ferry, thought to be derived from King Edward's Ferry. A superb engraving by Charles John Smith shows the village and the River Trent with windmills, sailing boats, rowing boats, houses and a cart at the water's edge. Three boats were kept constantly afloat, one for passengers, one for passengers and horses, and a third for carriages and droves of cattle. 'In fine weather, during the neap tides, the passage is made in a few minutes,' wrote Stonehouse. 'During spring tides and heavy freshes it is much more tedious and when the river is encumbered with ice, sometimes dangerous.' He said the place had recently (early nineteenth century) been called the town and port of Ferry on account of the number of vessels, keels and sea sloops which traded from there to Hull, Gainsborough and the west of Yorkshire.

He also wrote about the packet which sailed once a week to Hull. 'It was two days in making the voyage down and one tide in returning, sometimes two. On the voyage down the passengers stopped all night at Burton-Stather. The steam packets now accomplish the voyage in about four hours. Once a fortnight a person may embark at Ferry at half past nine, get to Hull about two, stay there until near four and return to Ferry again the same day by seven o'clock in the afternoon.'

In the summer of 1832 the village of West Ferry was attacked by Asiatic cholera. A man had died two miles away in Gunthorpe in April from the disease – 'he was a debauched character and had been drunk the day before' – and about the same time several deaths took place downstream at West Butterwick. 'Nothing more was heard of this formidable invader until the 10th of June when, the weather being wet and cold, the disorder again made its appearance in the house of one Joseph Waite, a labouring man, who, though he had made no previous application for relief, was found to be in a very destitute condition. Sarah, Joseph and Hannah Waite were all buried between June 11 and 15; the mother, very soon after lay-in, was safely delivered and recovered, escaping the disorder altogether. She was a great opium eater.'

A man named Keightly who had helped to carry some of the corpses, died and it was with great difficulty that anyone could be prevailed upon to carry him, or even lend a horse and cart for that purpose. After several deaths the inhabitants were very much alarmed and refused to go into the fields to work lest they be taken ill suddenly and die before they could get home. 'The church was opened for divine service occasionally in the evening and was very numerously attended. The public houses were entirely deserted,' wrote Stonehouse.

Many of the deaths were witnessed by Stonehouse and from an average of forty funerals in the parish each year, the number rose to seventy-three in 1832. Cholera next visited Gainsborough and then villages on either side of the river. East Ferry, opposite West Ferry, was free in June and July that year but was attacked in August. It was asserted by many people that when the tide began to flow up to Gainsborough it was accompanied by 'a very nauseous smell'. The disorder was confined to the riverside and no case occurred further than two hundred yards from it.

South Street runs with the river at Owston Ferry but a six-foot high concrete wall to keep out the river means that people living there cannot see the water from their ground-floor rooms. A potato mountain was growing in one farmyard, and an impress-ive new house contrasted sharply with the emptiness of an equally imposing but older property. Arches and alleys gave glimpses of the area behind the street and in grassy back streets the washing blew in the fitful breeze. I saw three attractive pubs, the Crooked Billet, the White Hart, and Red Lion and after staying with the river and watching the *Swinderby*, big and quite exciting, sail by I headed down North Street past Old Trent Hall.

Once out of the village I regained the bank, still offering its long views and providing good walking. I was forced back onto the road for a little while at Kelfield, passing a house with the date 1689 looking out over what looked like miles of ploughed fields.

Across the river, the village of Susworth has a neat look about it and the water here was wide enough for the driver of one red,

Door of the fifteenth-century church at Althorpe

unnamed coaster to leave his cab and attend to some matter at the side of the boat. Despite the flatness there is little to see but trees and pylons, ploughed fields and the occasional farm. The banking is easily fifteen feet up from the road, and over on the right Scunthorpe begins to appear a few miles to the north-east, spread out and with high-rise buildings.

Yet another pair of sister villages appear on opposite sides of the river: West Butterwick and East Butterwick. I went into the Ferry Boat Inn at West Butterwick, a depressingly deserted pub where the solitary drinker told me it must have been 'fifty or sixty years sin' packet ran to Hull on Market Day'. 'It started at Owston Ferry and picked up along t' way,' he said, and he thought it was forty years since the ferry ran.

Stonehouse said that from early times West Butterwick had had a Chapel of Ease which was once a strong and spacious building. 'The present building,' he said in 1839, 'bears very evident marks of having been erected out of the ruins of a more stately fabric.' There was also a small community of Baptists almost from the time they started in 1640. 'They have a meeting house with a burial ground adjoining and they generally baptise adults in the South Butterwick Drain, probably because one of the members of the society has a house upon the bank.'

In 1835, seventeen guineas were dug out of the grave in the West Butterwick chapel yard in which John Clarke had been interred about thirty years earlier. He was drowned in the River Trent as he was returning from Stockwith Fair with the sum of money on him. A captain who was sleeping on board a brig anchored off Kelfield dreamed he saw two men rob another man and throw him into the water and that the dead body 'had floated athwart his cable'. So strong was the impression that he went to look and sure enough, there was the body. The relatives, not finding on the body the money which he was known to have received at the fair, buried him in his clothes, as was usual when corpses had been some time in the water, and were convinced that the other part of the dream about the robbery and the murder must have been true also. Two men were picked up on suspicion and the captain of the vessel swore they were the men he had seen in his dream. As there was no other evidence they were discharged, but were still thought to have committed the crime until the money, which appeared to have escaped the search of the persons who prepared the corpse for interment, was discovered in his grave.

The Baptists, according to Stonehouse, exercised the power of ex-communication and recorded on June 27, 1706: 'We the church of God met at Butterwick. Whereas Isaac Lodson has taken a wife contrary to the law of God and gone to the Church of England, for such transgressions he is set apart by this assembly as a person not fit for com. with the church of God, until he shall purge himself of these evils.'

Just below West Butterwick I came across the biggest road I had seen in miles, a motorway running into Scunthorpe, about four miles away. The traffic seemed to be nearly all heavy goods and thundered over me on a modern bridge that is quite pleasing to the eye and which is one of the newest across the Trent, certainly the newest in the section I had walked by quite a few years. Two churches were coming up together, a small one set in the trees at Burringham on the east bank, and Althorpe with its tower on my side.

Before reaching Althorpe I had to go through Derrythorpe, once called Diddlethorpe, Deddythorpe and even Keddythorpe, where I saw a house with a notice offering Bramley apples at £1.30 per stone. How nice to see the old measures of weight still being used, although it took me a few minutes to do my sums and convert it to about 6 pence a pound.

As I walked into Althorpe, I saw another reminder of the Dutch drainage engineer of the seventeenth century: a block of bungalows called Vermuyden Court.

OS MAP 112
Information Centres: Gainsborough Library (0427 4780), Scunthorpe (0724 860161)

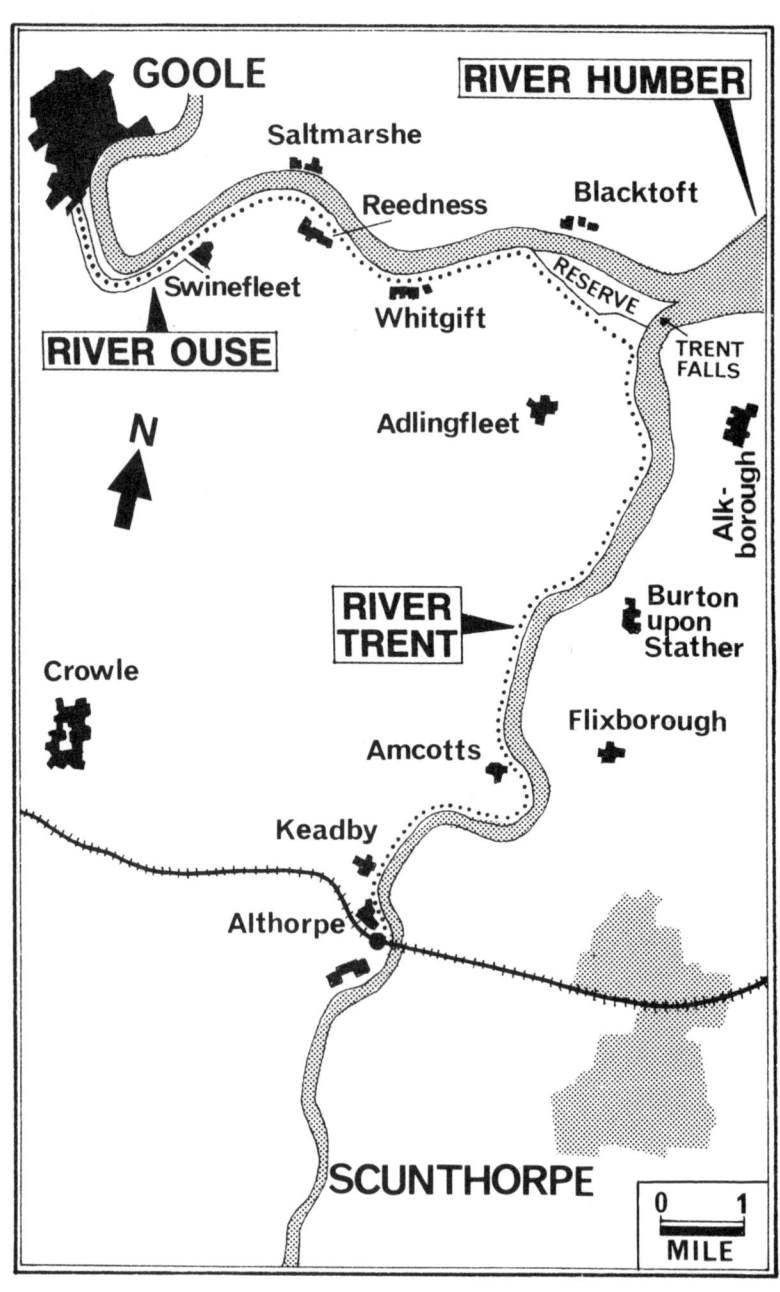

14

Althorpe to Goole

19 miles : 30.4 kilometres

The Trent is nearly finished. It has only about ten miles to run before joining up with the River Ouse to make the Humber which runs into the North Sea at Kingston upon Hull. The last bridge across the river is at Althorpe, a swing bridge with road and rail together, the railway line running from Doncaster to Scunthorpe and Grimsby. This, which had been Lincolnshire for a thousand years, is now Humberside. I had been out of Lincolnshire since just after West Stockwith, and even the east bank of the Trent had been Humberside since a couple of miles below the Butterwicks.

The fifteenth-century church at Althorpe stands right by the river, a church that saw restoration in 1868 when several coats of whitewash were removed from the bench of the sedilia and it was found to be a stone slab with the fine brass portrait of a fourteenth-century rector, William de Lound, wearing a flowing robe with four-leaved flower ornament on the collar. Althorpe was mentioned in Domesday Book with '. . . one carucate of land to be taxed. Land to one plough. Six sokemen have there one plough'. Stonehouse said there was a large old manor house, of the time of Queen Elizabeth, which had most probably been erected on the site of a more ancient dwelling. 'It is now much out of repair and, like everything of the sort in this country, has been used as a common farm house,' he declared with some despair.

The road took me away from the river and past the end of the road leading to Althorpe railway station and the King George V swing bridge, which was opened for railway and highway traffic on 21 May 1916. There was already a railway across the Trent – built in 1863 – but the new bridge put the two together on a bridge which last swung open in 1953. I regained the river by Keadby Locks, where there are three lanes of water on the left, a busy area with the Friendly Fox and the South Yorkshire Inn and a road naturally enough called Trentside. The Mariners Arms is just down the road, and again a five-foot-high wall stands between the river and the road. The mass of water around Keadby is its main feature and long provided the area with a good deal of work. In 1851 there were 602 inhabitants of Keadby, 188 of them working on ships and boats. Here, too, the Stainforth and Keadby Canal locks into the river and runs due west, where it links with the Sheffield and South Yorkshire Navigation. The Stainforth and Keadby was constructed in 1793 and is about thirteen miles long, running for most of its length close to the railway.

Across the River Trent from Keadby is the village of Gunness with its wharf, deriving its name from forming a ness or promontory in the Trent. Until the middle of last century it was not much smaller than Scunthorpe, about three miles away, but with the discovery of iron ore and the establishing of blast furnaces, the village of Scunthorpe turned into a town in no time at all. From a population of 303 in 1855 and 2,048 in 1881, it exploded into 70,000 by the middle of this century, absorbing five villages along the way.

From the bank by the River Trent I could see Scunthorpe to the east, with its towers and chimneys and high buildings, but I was heading north, past a house selling Trent Valley mushrooms, fresh daily at 80 pence a pound, and off into the country, into a part of England with little passing traffic which again is surrounded by water – the Trent, the Ouse and the drains.

The river is fast-flowing and wide; the fields are flat and ploughed, some of them sown; farms are dotted about, and the inevitable pylons stride off into the sun, which that day was low and watery and reluctant to smile. Cranes clattered and banged and squeaked on the other side of the river which bends sharply and heads due east for a time. There were plenty of ships lining the bank, the *Vanernsee* and the *Stevonia*, and one from Limassol called *Simone*. The river swings back again to make sure it does not miss Amcotts on its way to the Humber. Amcotts was named after John de Amcotts, who was returned to Parliament as a member for the city of Lincoln in the reign of Edward II, and it saw a considerable change in the course of the river in the eighteenth century.

The poorer people around here used peat for fuel for many

years and sometimes dug as deep as six feet. In 1747 one peat-digger on the moor came across a human skeleton in an upright position. George Stovin of Hirst, who wrote about the discovery in the *Gentleman's Magazine*, looked in vain for a coin to date the skeleton. He took away the bones of a hand and both feet, which were entire 'and in their natural order, enclosed in shoes or sandals'. The leather of the shoes appeared firm and neat and very little, if at all, affected by the moist conditions they had probably been in for centuries. One shoe was sent to the Royal Society which dated it about the time of the Norman Conquest and identified the skeleton as that of a woman. Shoes of that particular description were first worn around the time of Edward I or Henry III. It was thought, from the skeleton having been found in an upright position, that the unfortunate woman had perished accidentally. 'She had probably lost her way and sinking into the soft ground of those desolate moors was unable to extricate herself,' wrote Stonehouse. 'The state in which the skin was found, being tough and stretching like a piece of doe leather, is easily accounted for. It had been tanned by the moor water. Mr. Stovin informs us that he buried the remains of the lady in Amcotts chapel yard.'

A road which starts near the river swings into Amcotts, a plain village with a war memorial on which are the names of three men called Walker, brothers perhaps, and all in the 21st Battalion of the King's Royal Rifles, two of whom died from their wounds some years after the war. Private Alexander M. Walker was killed on 15 September 1916, Private Edward Walker died on 23 May 1921 and Rfn. Ernest Walker on 10 April 1927, both from wounds received in France. Even here, in this quiet, out-of-the-way village, the youngsters had been up to their tricks, turning a road-sign upside down and facing the sign bearing the name Amcotts into the village and upside down. This is the land of the potato and the parsnip, and on a road that sees little traffic a huge lorry suddenly appeared, stacked high with the vegetables. The land is still protected from the Trent by the high banking, and running into the river from the fertile country comes the Paupers' Drain.

Across the river, through the mist, I could just make out a few trees and the occasional farm, and in there somewhere was the village of Flixborough which was struck by disaster one Saturday

Greenshank

Little Grebe

Bearded Tit

Cormorant

Merlin

Bittern

Blacktoft Sands Nature Reserve stands where the Rivers Trent
and Ouse join to make the Humber. Illustrated are several of
the birds which have been spotted there

afternoon in the summer of 1974 when a chemical works blew up and twenty-eight men were killed. It was the biggest explosion Britain had seen since the Second World War. It wrecked the Nypro works, and not a house in the village escaped: tiles were torn from roofs, windows were shattered and life came almost to a standstill. Yet none of the dead came from Flixborough, with its population of 300. Most of the workers killed had come from Scunthorpe, and many more from the surrounding villages. Seven of the villages were evacuated – Roxby, Burton upon Stather, Amcotts, Normanby, Gunness, Wintringham and, of course, Flixborough – in case of another explosion. The works were rebuilt over five years at a cost of more than £30 million but were closed down in 1981 because of a recession in the man-made fibre trade. Until the explosion Flixborough had lived a quiet life since being inhabited by Danish invaders more than a thousand years ago. The village was recorded in Domesday Book under its Danish name, Flichesburg, and was the birthplace of Sir Edmund Anderson, a judge who took part in the trials of Mary, Queen of Scots, and Sir Walter Raleigh in the sixteenth century.

The spire of Luddington church stands out as a landmark over the flat emptiness, set among trees where it has been since 1855. The road, which has kept the ever-widening Trent company since Amcotts, branches off after a couple of miles and heads for Garthorpe where, according to Stonehouse: 'The Red Lion staring o'er the way, Invites each passing traveller to pay.' There was a market at Garthorpe, probably frequented chiefly by people who arrived in boats.

One of today's boats, a coaster from Hull, chugged upstream against the tide and the breeze, and across the river I could make out the Burton upon Stather wharf, stacked with timber, and a cluster of houses. But the mist hid any sight of the town up on the cliffs, which are wooded in parts and form the highest land I had seen for many miles. In contrast to the cliffs on the east bank, the west side of the river is still made up of sowed fields stretching into the distance.

Burton upon Stather is a picturesque village with some charming cottages, built, so it seems, out of a product called clunch, a form of chalk. Among the monuments in St Andrew's Church, which is twelfth century, are an early fourteenth-century cross-legged knight from the Crusades who wears chain armour and who was once at Owston Ferry, and that of Sir Charles Sheffield (1776) whose family owned much land in the area. Normanby Hall, which stands nearby, was designed by Sir Robert Smirke in 1820 for Sir Robert Sheffield, replacing a similar Georgian house.

Burton upon Stather took its name from the staith or mud which gathers here at the rising of the tide. It was once considered the metropolis of the Trent and in the reign of Edward II was awarded a charter for a weekly market and two annual fairs, one to last fourteen days in May, the other four days from the eve of Holy Trinity. In those days large vessels were unable to proceed any great distance up the river without danger, but once trade started to increase in Gainsborough, the prosperity of Burton declined. In 1770 the banks of the river gave way a little below Gainsborough, and in a few days all the low land around Burton was flooded. The shores on each side of the river were then secured by a large number of jetties. Seven years later a vessel carrying gunpowder caught fire and exploded with a noise that was heard for miles. Houses at Burton were unroofed and it was estimated that £3,000 worth of damage was done to the church and other buildings. Happily, the sailors had escaped from the ship before it blew up.

In the churchyard, over the body of James Scott, once a sculptor in Burton, are the words:

> Praises on tombs are vainly spent,
> A good name is a monument.

Another inscription over somebody called Roberts reads:

> Our life is but a winter's day,
> Some only breakfast and away,
> Others to dinner stay and are full fed,
> The oldest man but sups and goes to bed,
> Large is his debt who lingers out the day,
> Who goes the first has the least sum to pay.

I had read of the magnificent three-mile walk along the clifftops from Burton to Alkborough, with its marvellous views over to where the Trent and the Ouse combine to form the Humber. Some weeks later, after a visit to Beverley, I travelled over the

Humber Bridge to Burton upon Stather and walked along the cliffs. The view is quite breathtaking in parts, especially where the Humber is seen for the first time. It is almost like looking out to sea, a truly impressive spectacle.

Alkborough is an ancient village, and William Andrew said of it in 1836: 'Too much cannot be said of Alkborough. Here is a scenery on which an eye may rest with rapture. Beautiful.' Here is a medieval turf maze cut in the ground known as Julian's Bower at a spot that was once a place of fortification as well as amusement. An iron copy of it is let into the stone floor of the church porch. One of the books on Lincolnshire said Alkborough had a Yorkshire bleakness about it. Now it is Humberside, though not yet Yorkshire. But it is getting close.

A few miles east towards the sea is Barton upon Humber, the most important port on the river until Hull was founded in the reign of Edward I. Daniel Defoe described Barton as '. . . a town noted for nothing that I know of, but an ill-favoured dangerous passage, or ferry, over the Humber to Hull; where in an open boat, in which we had about fifteen horses, and ten or twelve cows, mingled with about seventeen or eighteen passengers, called Christians; we were about four hours tossed about on the Humber, before we could get into the harbour at Hull; whether I was sea-sick or not, is not worth notice, but that we were all sick of the passage, any one may suppose, and particularly I was so uneasy at it, that I chose to go round by York, rather than return to Barton, at least for that time.' Now the Humber Bridge, built in 1981, crosses near here, and a fine viewing area with a car-park has been constructed close to the river.

The Trent is almost spent, but as it nears its end it broadens to more than a quarter of a mile wide, with mudbanks showing. I stayed with the river, alone with the water and the birds wheeling and calling over the mudbanks. I went over and through gates and past an isolated farm that must hardly see a traveller in this cut-off corner of England close to the joining of the rivers.

Over to the left is Adlingfleet with its thirteenth-century church and brick houses, originally named Aethelingsfleet from Edgar Aetheling, heir to Harold and the crown of England. He accompanied the Danes on their expedition in 1068 when the sons of King Sweyne, his brother Osbeorn and five other Danish chiefs of high rank entered the Humber as allies of the Saxons who, north of the Humber, were attempting to escape from the Norman suppression. As winter approached, the Danish ships moored at Trent Ness between the confluence of the Ouse and Trent and just within the channel of the River Don. Their camp was in a strong position so that few forces could defend it. But a large bribe from William the Conqueror took the Danes away without fighting – their last excursion into England.

Trent Falls is the rather grand name for some rapids, but I was unable to see them as the path turns left before the river swings right in its last few hundred yards. The path skirts Blacktoft Sands Nature Reserve and crosses Adlingfleet Drain, staying on the floodbank leading away from the river. The nature reserve is between the river and the bank, with masses of six-foot-high grass swaying in the stiff breeze. I could make out the lighthouse at the joining of the Trent and Ouse, and in the distance I saw rise a huge flock of birds I could not identify. A Royal Society for the Protection of Birds' sign asked me to walk below the bank to avoid disturbance – so down I went.

Blacktoft village with its church tower came into sight across the water – which I still could not see – and here at Blacktoft jetty, I understand, is where the boats tie up when they have got the tide wrong. Here they can moor temporarily so they do not ground, a long practice that led to the local Customs officer reporting in 1888 that the pier was a place where '. . . grave mischief might possibly break out at any time.' He pointed out that steamers inward bound from 'dangerous continental ports' often stayed at the pier during the night, and vessels bound for the sea, with stores from Bond on board, were also often obliged to stay there for the same reasons. The dangers of avoiding Customs were all too clear, and from 1888 to 1892 a permanent Customs officer stayed there.

I continued walking along the edge of the nature reserve. Inside are huts, and a man with binoculars shut the door of one close to where signs pointed to Xerox, Ousefleet, First, Townend and Reedbed Hides. I soon reached the reception hut which lists birds that have been recently spotted. Among them were the red-throated diver, Leach's Petrel, cormorant, bittern, whooper swan, pink-footed goose, teal (with a peak count of 710 which might have been the flock I saw), pintail, shoveller, tufted duck, goldeneye, merlin, peregrine, golden plover (up to 850), curlew sandpiper, ruff, spotted redshank, short-eared owl, stonechat, greenshank and green sandpiper. The birds that had been seen that day included hen harrier, merlin (hunting the lagoons and reed beds), 500 teal, little grebe, spotted redshank, water pipit and bearded tits, small numbers of which gathered at the edge of the reed beds in the mornings.

I left the reserve at the hut and took the road to the village of Ousefleet, passing – at the reserve entrance, in an area with picnic tables – a seat put there in memory of Geoff Coe (1927–83) RSPB, Airedale Group. Ousefleet straggles along the road and I soon found the lane that took me to the bank of the Ouse. It was lovely to see another river after more than ninety miles alongside the Trent. The Ouse is impressive here, about 400 yards wide, tidal and deep enough to carry huge ships as far as Selby and Goole. Across the water I could make out another of those windmills without arms, at the village of Yokefleet, and up ahead was one of the many lighthouses, which, along with beacons, dot the river for miles.

The Ouse rises on the borders of Yorkshire and what was Westmorland. Its most northern branch is the Swale, and the river runs by Reeth, Richmond and Catterick. The Bedale river joins just below Morton Bridge, joined by the Codbeck and Ure, and at Linton becomes the Ouse. The city of York took over responsibility for the Ouse in 1462 and made the river navigable to Swale Nab, sixty miles from where the Ouse joins the Trent, and beyond. The river is tidal to Naburn Lock, about five miles below York, which still sees commercial traffic.

The path stays with the river and passes along the end of a garden with heavily laden apple trees. It joins up with the road from Ousefleet, which has been closing in all the way, to the charming village of Whitgift with its jetty, ancient brick manor house and church of St Mary Magdalene, built in 1304. In the graveyard is a stone to the memory of John Cuthbert who was drowned at Hull on 24 December 1852, in his twenty-second year. Alongside one another are identically carved stones, one to Hannah, the other to Eliza, both daughters of Benjamin Wilkinson and Elizabeth Martin of Reedness. Both died aged twenty-one, Hannah in 1865, Eliza in 1861.

Whitgift was once the most important ferry for the lower reaches of the river. The water is wide here, and the tide can be rough – on 3 December 1614 Sir John Sheffield (son of the then

The jetty at Whitgift on the River Ouse

president of the Council of the North), two brothers and servants were drowned when attempting to cross. The poet Michael Drayton wrote an elegy about the tragedy but talked of the Humber not the Ouse. King Charles I used the ferry when going from York to Nottingham to raise his standard in 1642, and John Wesley probably used it frequently. Wesley, who was born in Epworth, near Owston Ferry, seems to have had an unpleasant passage in May 1753: 'I preached at Pocklington again and rode on to Whitgift Ferry. It rained a great part of the way and just as we got to the water a furious shower began which continued about half an hour while we were striving to get John Haime's horse into the boat; but we were forced after all to leave him behind.' The ferry landing has gone but the ferry house remains as the village inn.

I walked along the grassy floodbank, with its grandstand view, which separates the road and houses from the river. The two part company again once they have gone through Little Reedness, the river narrowing here to about 250 yards wide. A yellow arrow led me across a bridge over a brook running into the Ouse which leaves the village of Reedness on the left and Saltmarshe, with its houses and farms by the river, on the outside of the bend on the right. A hall stands impressively on the bank, well protected from the river, it seems, by a stone embankment.

The path continues along the inside of the gentle bend, and the British Transport Docks Board announce on a sign that it is their land, and they are allowing me access. A barge from Hull, lying low in the water, was going upstream just as the path started to go close to the river. A sign told me to keep to the path, which had become muddy, but metal gates soon pointed me away from the river bank to join the road at Swinefleet, an old-fashioned village with character. The name of the village came from Sweyne, the Danish King, and is Sweyne's harbour or naval station. The road through the village, which soon meets up with the main road running from Goole to Crowle, is called the King's Causeway. On this road, about two miles from Swinefleet, is a field that was known as Ned Mangarell. Tradition says that about 1690 a man of this name was hanged here for murder, and a portion of the gibbet on which he was executed was said still to be standing in 1927.

I walked down Swinefleet's High Street with its modern church and the King's Head back to the river bank and its beacons. The road keeps with the river as it performs another U-turn, but I stayed on the bank and looked across the water to Goole with its cranes and church spires. The tall grass that had been so evident around Trent Falls was back again, hiding the River Ouse from view as I passed Goole Hall and started the run into Goole. As the road swings left, the path stays with the river behind houses and farms for a few hundred yards before running left alongside Cochrane Shipbuilders to rejoin the road into town.

The road goes through Old Goole – the original village before the present town took over – and crosses the Dutch River with the pub New Bridge on one side and the Vermuyden on the other. Vermuyden (whom we met at Axholme), was the nephew of the Dutch ambassador to the Court of Charles I and built embankments to prevent the River Don from flooding. Unfortunately the plan went wrong and a disastrous flood was the result, forcing him to make a new river-bed from Snaith to Goole. This was the Dutch River.

OS MAP 112
Information Centres: Scunthorpe (0724 860161), Goole (0405 2187)

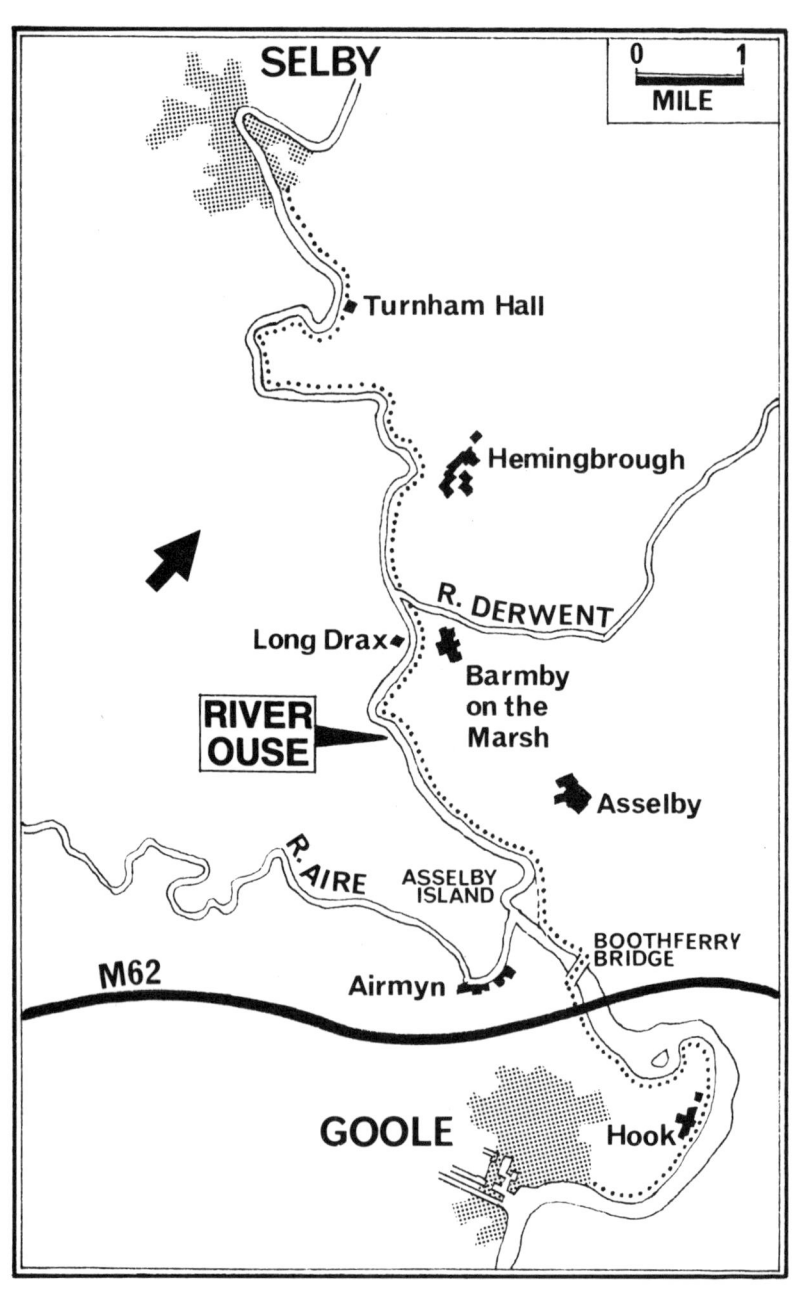

SELBY

0   1
MILE

Turnham Hall

Hemingbrough

R. DERWENT

Long Drax

Barmby
on the
Marsh

RIVER
OUSE

Asselby

R. AIRE

ASSELBY
ISLAND

BOOTHFERRY
BRIDGE

M62

Airmyn

GOOLE

Hook

15

Goole to Selby

16 miles : 25.6 kilometres

For somebody like me, who had never before been to Goole, the size of its port and the number of ships that lodge there are quite a surprise. The population is around 18,000, and the town is fifty miles from the sea.

If anything, I suppose, Goole should be a dozy market town but, instead of a cobbled square or an ancient market-place, there is the prow of a huge ship towering over you or a crane searching the depths of a hold for something to clutch and dangle in the air. Consequently its sounds are those not of a country town but of a busy port: the screech of the gulls, the noise of the ship's siren, the whirr and rattle of the cranes. A gatekeeper stands by to supervise the swinging of the road bridge to let some boat go home to Copenhagen or Cadiz. But this is still the heart of Yorkshire, where a sandwich and a cake and a pot of tea will not bankrupt you. The market hall is still true to northern tradition, and the pubs are down to earth. It is not an old town – in fact, it has only one building dating from before 1820 – yet it has a settled warmth about it, a comfortable place to be in.

Until the 1820s Goole was just a hamlet on the Ouse with a population not yet reaching 500. The Victorian age was almost upon England when the proprietors of the Aire and Calder Navigation decided to construct a new outlet to the Ouse for their canal and river system. The existing Selby Canal was inadequate, and the deeper Knottingley to Goole canal replaced it and was opened in 1826. The population exploded as the port developed, and the new town soon dwarfed the hamlet of what is now Old Goole on the Swinefleet road. A church was built by the men who controlled the Navigation; a market and a school were established; the port blossomed and has today still managed to hold on to its fairly considerable trade while much more sizeable and seemingly indestructible ports such as Manchester and Preston have withered.

A man called Bartholomew was resident engineer of the Aire and Calder and was responsible for the introduction of something new in water transport in the form of steel 'compartment' boats for carrying coal, twenty feet long, which were attached together like railway waggons to the back of the tug. Hydraulic hoists lifted the 'compartments' out of the water, and the contents were poured into the ships. As many as thirty could travel together, and they were known as 'Tom Puddings', which could have something to do with the Yorkshire pudding tins. They still exist today but to my great disappointment I missed the sight of them on the Ouse. The closeness of the M62 has, perhaps, helped Goole maintain its position as England's farthest port inland.

I walked through the town past the docks, the railway station and the bustling shops, and on past the Market Hall and the library, down North Street to the river, impressively, marvellously wide here, 250 to 300 yards across. The banking, which had provided me with wonderful walking for miles and miles was still there as the Ouse, which was still on my right, runs parallel with a quiet road which has a house named Harbour House. There is a well-laid path running out of the town towards the fields, the trees and the good Yorkshire countryside. Behind me lay the cranes and the spire of the church, and I could make out a ship in dock. The river performs a double U-turn around Goole, with the town lying within the upstream U as the river swings sharply from heading due east to heading south almost to encircle the town with the help of Dutch River and the River Aire. The path gives way to the familiar grassy bank and runs along the bottom of the gardens of some quite impressive houses. Once past the cemetery I could see the massive railway swing bridge, built in 1869, which takes trains from Hull to Doncaster and Wakefield and Leeds. The river is over 250 yards wide here and the bridge, named on the Ordnance Survey map as Goole Bridge but also known as Hook or Skelton Bridge, is about 280 yards long.

I moved on past a farm bursting with pigs in an area that is still residential as I approached the village of Hook, where property touches the river bank. Several of the houses have living-rooms upstairs so people can see over the banks to the river and beyond.

A monastery is said to have stood in the fields close to the church at Hook, an ancient church with a window showing Queen Victoria visiting the wounded of the Boer War. Hook Hall is set among the trees, tight in the crook of the river as it turns back on itself. Across the river are the wharves, cranes and ships of Howden Dyke, where, I was told, there used to be a sugar factory. There used also to be a ferry here, from as long ago as the thirteenth century, at one of the narrower parts of the river.

Howden Dyke stands at the top of the great loop as the river turns right round and heads south again, back towards the cranes and church spire of Goole for a few hundred yards. The river really does resemble a hook here, although I guess there will be a few educated reasons for suggesting some other way of arriving at the name of the village of Hook. I walked round the inside of the loop and there, in front, was the mighty M62 motorway, close to Boothferry Bridge, which was formerly the only road bridge across the Ouse between Selby and its union with the Trent twenty-four miles away. The great highway gracefully rises high above the river on about thirty two-legged columns, a giant centipede, elegant and motionless. The river splits here around a large island, and the path took me past Westfield Hospital, which was closed, and gave me a view of Howden's imposing church tower beyond the motorway. As I approached the motorway, a barge from Hull passed beneath it, heading towards Goole, and the skipper looked up at the road above him, maybe marvelling at the great feat of engineering. The bank lifts the road onto its legs, which grow longer and longer, with two sets of them resting on islands in the river.

A road linking Hook with the A614 trunk road across the neighbouring Boothferry Bridge runs alongside the river, below the banking, which still provides the best place for the walker. The colossal iron Boothferry swing bridge is only about half a mile upstream from the motorway. It was opened in 1929 and replaced the longest surviving major ferry on the river. Now it sits back and watches the majesty of a motorway which links Hull with Liverpool.

A Humberside County Council employee working on the bridge pointed out the driftwood at the foot of the floodbank and showed me the handrail where a four-metre tide reached. 'A six-metre tide would get to the houses, but I haven't seen that,' he said. I asked how many times the bridge would need to swing

New Bridge Hotel which stands near Dutch River at Goole

each day. 'Some days we'll get five, the next day maybe none,' he said. 'We've just had a barge through on high tide and we had to swing for that.'

I had to cross the river here and walk with the Ouse on my left for the first time. The River Aire was little more than half a mile ahead on the left, and if I had stayed on the southern bank I would have had to walk miles before I could cross it. The River Derwent runs into the Ouse from the right about three miles ahead, but at least the detour was not as long, probably about three miles. I left Boothferry Bridge behind me but the path and the bank soon leave the Ouse to get round Asselby Island, where the channel has altered over the centuries and where the island disappears, then reappears. Because of the island, I missed the sight of the Aire, one of Yorkshire's many lovely rivers, pouring into the Ouse.

The Aire runs seventy miles after springing from Malham Moor, and its early life is in picturesque surroundings. It provides such industrial centres as Leeds and Bradford with attractive scenery on their doorsteps and is a natural dividing line between the industry to the south and the scenic country to the north.

By the time I had regained the Ouse, eight minutes after leaving it to bypass Asselby Island, the Aire was well behind me and the Ouse was running through open countryside with little life on either bank. It was flat and open for two miles, with few sounds except the ripple of the water, still tidal and now heading back to the sea, and in the distance the faint sound of the traffic on the motorway. Over to the right, Asselby village is strung out along the road, and from my vantage-point on the floodbank I soon got my first sight of Drax Power Station with its cooling towers dominating the horizon. Rusholme Grange stands by the water on the other side of the river, deep in Yorkshire country that is isolated and peaceful. The river swings sharp right away from the power station and heads for Barmby on the Marsh with its church with a green, round-topped tower.

Before reaching the Derwent I passed the site of the old swing bridge which, from 1855, carried the Hull and Barnsley railway across the river until the railway was closed. The bridge stayed there for many years before it was removed, and today the banking still approaches the river before coming to a halt on its huge forty-foot-wide stone supports near to the water. The line has been overgrown with gorse, and a fence has been erected six yards from the end of the huge stone buttress which stares across the water . . . at its double.

The elegant, slim line of Hemingbrough church spire shows up ahead, and a sign on the river bank tells of the impending arrival of the River Derwent 250 metres ahead. I must have had an old Ordnance Survey map, because there was no sign of a bridge across the Derwent on mine and I looked to be faced with a three-mile detour just to get to the other side and regain the bank of the Ouse. It came as a pleasant surprise to find a manned control tower at the entrance to the Derwent and a bridge across it which saved at least an hour's walking. I was told the building was there to control the flow of water into the Ouse as well as attend to the lock into the river for craft. The man in charge thought the bridge, which provided a public footpath across the Derwent, had been erected in 1974. I looked at my map from the library – 'Crown Copyright 1974'!

The man in the control tower, who had a dog to keep him company, told me there used to be a ferry at that spot, from a house which he pointed out to me to the pub across the river. He reckoned the railway bridge had been taken away about five years earlier. 'Cut us off from the village of Long Drax over there,' he complained. 'Now it is eighteen miles to get there.' I looked at my old map again. It certainly was a tortuous way round. We were exactly halfway between Goole and Selby, at a part of the river which is now only about sixty yards wide. The countryside is perfect. The path and the walking are lovely as the river snakes its way towards and then away from the village of Hemingbrough.

There used to be a ferry here at Newhay, close to Hemingbrough, although it is hard now to see why, as there is nothing to be seen on the other side of the river. Hemingbrough is, however, an ancient place. It was here before Domesday, known then as Hamiburg, and with a tradition that the Romans had a fort here. This area, between the Ouse and the Derwent, has long been – and, I suppose, still is – agricultural and thinly populated. The river used to pass close to Hemingbrough, and the Ordnance Survey map shows the old course of the river, which also passed close to the neighbouring village of Cliffe.

G. Bernard Wood, in his book on Yorkshire villages, said that in the triangular tract of country between Castleford, Goole and Selby '. . . few villages need detain us, and those few largely on account of their churches.' Drax, which has a Norman church and once had an Augustinian priory and a castle, gets a mention, and so too does the beautiful spire of Hemingbrough's church of St Mary, 180 feet high – 'fit to grace a cathedral'. Wood also writes of the carvings, including two-headed dragons. The church, built of the same fine limestone as that used in the construction of York Minster, originally stood on the bank of the river in a position which commanded a considerable view. 'The picturesqueness, as well as the usefulness of the village, has been considerably lessened by the withdrawal of the Ouse to a fresh channel,' wrote Thomas Burton in 1888.

The spire, which was known as the Hemingbrough broach – defined as an octagonal spire carried on a square tower which has no parapets – is one of the landmarks of the country and can be seen for miles in almost every direction, but it is at its most impressive from close to, and particularly across the fields from the bump in the river about six or seven hundred yards from the church. It looks huge, dominating the scene, its white stone shining. On a second bump in the river are Goule Hall Farm and Goule Hall Pumping Station, erected in 1959.

The river here is full of turns, winding in all directions, with Cliffe to the front, the right and the back, all within minutes. Another power station shows up in the distance, that at Egg-borough beyond the River Aire about six miles away, to keep company with the one at Drax, now on the left and still not all that far away as the river desperately tries to find its way out of Selby.

The river bank had been freshly mown, and the path provided more good walking alongside a river that was now sweeping back to the sea and leaving the mudbanks glistening. I noticed that the beacons across the river were named – one of them after a pub, the Brown Cow, and another, provocatively, Thief Lane. The towers of Selby Abbey were directly ahead, perhaps four miles away, before the river decided to turn away again, putting Selby behind me, Drax Power Station to my right, instead of left, and Hemingbrough church ahead of me! The river had turned right round and quickly had to do another U-turn to resume its correct course towards the abbey. And on the outside of another bend was Turnham Hall, once the home of Thomas Burton.

Selby Abbey, which dates from the time of the Normans

Turnham Hall dates from the time of Richard I, a timber house covered with tiles. It was rebuilt at the end of the sixteenth or beginning of the seventeenth century, pulled down and rebuilt by William Burton between 1796 and 1802. During this rebuilding, all traces of the moats by which the original house had been protected, disappeared. Hedges of box or yew and stately elms were for a long time the roosting place of a colony of peacocks, but in the last hours of 1778 'a furious tempest raged' and the elms and peacocks were destroyed. Thomas Burton, who wrote *The History of Hemingbrough*, was born here, was baptized at the church in 1801 and died in the house in 1883.

Soon after leaving Turnham Hall, I saw Selby Abbey directly ahead again, and this time the river path was straight and true as it headed for it. The path was still good and the grass, which had been cut, was being burned off. Soon after, I passed several houses by the water, some of them empty and forlorn. Cherry Orchard Farm appeared on the right, industry on the left, and then came the surprising sight of a Sealink British Ferries boat, *St Cecilia*, from London, being fitted out. The ship was a hive of activity with oxyacetylene cutters going, cranes swinging and men banging and hammering to their hearts' content. Just beyond it was the abbey, a wonderful building.

I passed the unimaginatively named Bleak House just before reaching the railway swing bridge and the sight of ships from Elsfleth and Rendsburg. A train crossed the swing bridge and pulled into Selby station, a hundred yards away, and as soon as it had gone, one of the ships, named *Gesche*, hooted impatiently and the bridge swung open to allow it through. The ship had half the river to use, a delicate operation as it backed through the gap between the watchtower and the stone block holding the bridge in mid-river. I stood and watched for a few minutes while the other ship unloaded a cargo of Wafolin, a binding agent for animal food pellets, before I took a closer look at the abbey, one of the most glorious sights of the entire walk.

OS MAP 105
Information Centres: Goole (0405 2187), Selby (0757 703263)

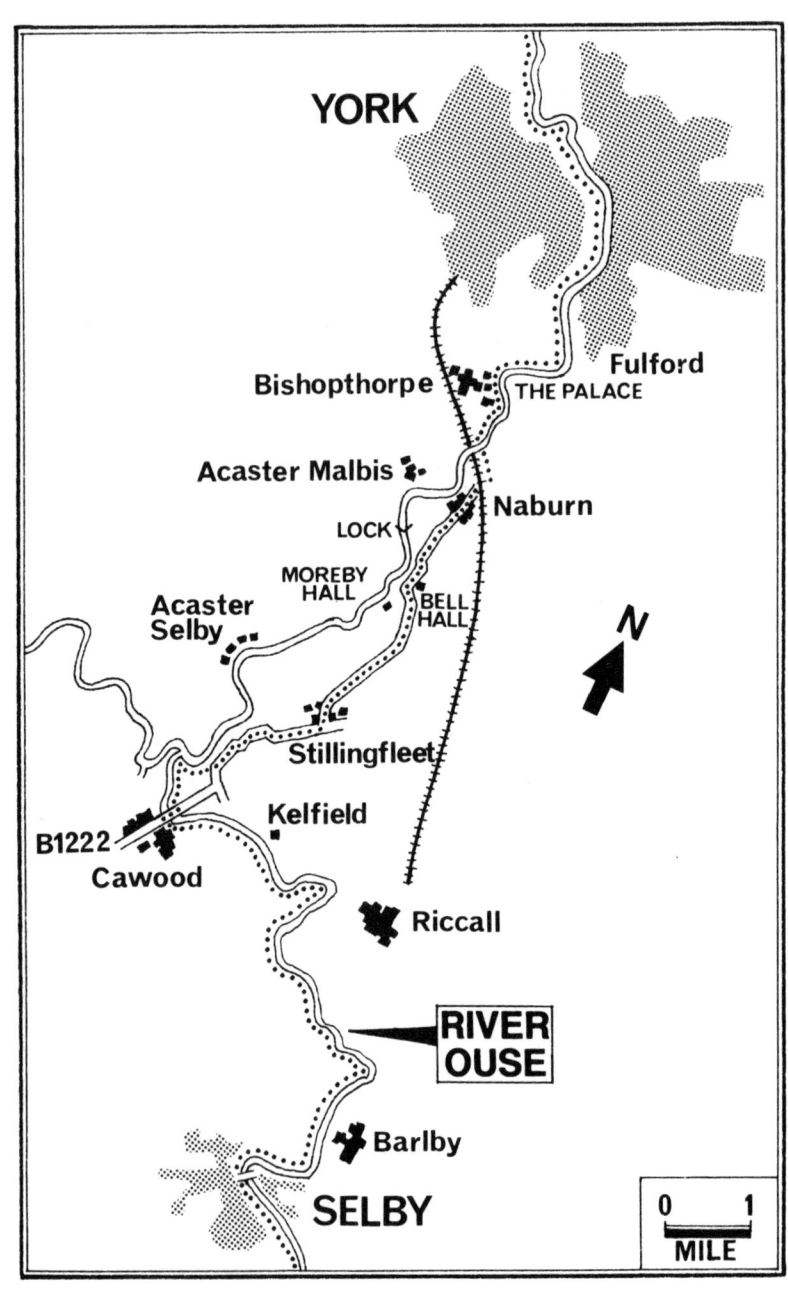

YORK

Bishopthorpe
THE PALACE
Fulford

Acaster Malbis

LOCK
Naburn

MOREBY
HALL
BELL
HALL

Acaster
Selby

N

Stillingfleet

B1222
Kelfield

Cawood

Riccall

RIVER
OUSE

Barlby

SELBY

0   1
MILE

16

Selby to York

19 miles : 30.4 kilometres

James Mountain wrote in 1800 that the town of Selby had improved from being a small fisher-town to a place of considerable business by its situation on the Ouse and the branches of inland navigation forming a junction with it.

A much later writer, Baron F. Duckham, referred to this period, between 1778 and 1826, as Selby's heyday, when the Selby Canal was the key to the Aire and Calder Navigation: 'The shipbuilding industry grew, rope and sail-making flourished and a branch custom house was established,' he wrote. J. Douglas Porteous said that, as an exporter of inland coal, Selby had been able to oust the sea-coal trade from the upper Ouse. The population increased from 2,861 in 1801 to 4,097 in 1821, one third of the working population being employed directly in the waterborne trades, '. . . a dependence which was Selby's undoing. Like Ulverston, Selby was finally eclipsed by the railway and also by the canal which created the port of Goole.'

Daniel Defoe had been in the area about a hundred years earlier, when he linked Selby and Howden as towns of good trade, '. . . the first being seated where the Ouse is navigable for large vessels, has a good share in the shipping of the river, and some merchants live and thrive here.'

Selby is a grand old market town, its character springing from its beautiful white abbey, whose origin is contained in a delightful legend which goes back to about the time of the Norman Conquest. It seems that a monk, Benedict, who was not too highly thought-of by his fellow monks in the monastery of Autun in France, claimed St Germanus had appeared to him in a vision and warned him of approaching dangers from his fellow monks. He fled from the monastery and took with him one of the most sacred possessions of the house, the finger of St Germanus, as a precious relic and talisman. He had been told to go to 'Selebaie' in England and, feeling that Salisbury was not the right place, sailed round the east coast to the mouth of the Humber, went up the river and came to a river bay which was frequented by seals and was called Sealby. It has also been said that his dream was fulfilled by three swans landing in the water at the same place. Whichever it was, he landed and, under a large and spreading oak near the riverside, erected a large cross. He put the treasured finger alongside it and spent much time in devotion. The cross was noticed by Hugh, the Norman sheriff of the county, as he was going up the river and, after hearing the monk's story, he gave Benedict his own tent as a tabernacle for the precious finger of St Germanus.

From this beginning sprang Selby Abbey. William the Conqueror himself became interested and visited the place in 1070, providing land for the site and endowment of the abbey. Wooden cells were first built and then the extensive domestic edifices. The sheriff's tent was replaced by a worthy church, a portion of which still forms part of the abbey, which is dedicated to St Mary and St Germanus.

Benedict became the first of more than thirty abbots, and his abbey has withstood one or two impending disasters, notably the fire which swept through it in the fourteenth century, the collapse of the central tower in the seventeenth and another fire in 1906 which badly damaged the fabric. Benedict is portrayed in one of the windows, and in another, in forty-six scenes, is the story of St Germanus, who was born in Auxerre in the fourth century. It is a beautiful church, facing the market place and set behind railings. A saxton called John Archer is remembered in verse in the nave:

> Near to this Stone lies Archer (John)
> Late Saxton (I aver)
> Who without Tears thirty four years,
> Did Carcases inter.
>
> But Death at last for his works past,
> Unto him thus did say,
> Leave off thy Trade be not afraid,
> But forthwith come away;

Without reply or asking why,
The summons he obey'd,
In seventeen hundred and sixty eight
Resign'd his Life and Spade.
Died September 15th AE 74.

In the abbey church is the coat of arms of the Washington family, three stars and two stripes in stained glass. It was certainly there in 1584, and perhaps two centuries earlier. John Washington emigrated to America in 1657, and it was his great-grandson who became the first president of the United States.

The monks established a ferry in the eleventh century, and it ran until 1791, when the first bridge was erected. The bridge was sorely needed because the '. . . danger, difficulty and at some-times, impossibility, in getting over the river by ferry boats, rendered it extremely hazardous'. In one month, from 12 September to 12 October 1790, it was said that 8,743 people went across the ferry on foot, 3,052 more with horses, and in addition animals including 2,248 sheep, 127 oxen and 66 hogs. One coach and fifteen chaises went across, as well as nineteen waggons and carts. That bridge, made almost entirely of timber, stood there until 1971, when it was replaced by the present steel one.

Today it is still a toll bridge, charging 7 pence for cars when I was there, and 6 pence a ton for commercial vehicles – causing the most terrible havoc and frustration as traffic built up waiting to cross the river. A notice proclaims rather grandly: 'The Company of Proprietors of Selby Bridge', and standing against the wall are two of the wheels, one small, the other eight feet high, from the original manual gear which was used to swing the bridge for shipping. 'They were in constant use from 1791 to 1970 when the bridge was reconstructed and mechanised swinging was intro-duced,' said the notice.

I was a little undecided as to which side of the river I should take, but I was soon put right by a man who worked in a nearby flourmill. I crossed the toll bridge, so the water was on my right, and went past Allied Mills and Ideal Flour, and past another mill and offices on the right before returning to the bank. The huge mill of Rank Hovis was across the water, and barges lay at rest, with names like *Selby Martin*, *Phillipa* and *Margaret*, *Leo* and *Oraco*,

Cottages at Cawood alongside the River Ouse

while *Elizabeth* and *Charles* sat quietly next to one another. Once again the path was good, ideal for walking, with contrasting views on either side of the Ouse. To the left were the fields and the flatness, to the right factories and barges, and ahead was the village of Barlby with its own mills. The tide looked to be just about at its highest, washing over the bank, and trees were standing in the water, which was still wide, somewhere around eighty yards. The abbey stayed in sight for some miles, a prominent landmark between the flourmills, and I could hear a clock striking across the fields.

The path is over stiles and through gates, and the river does a sharp U-turn – more a V-turn – as it makes a shape like a pagoda and almost touches the A19 trunk road between Selby and York at Turnhead Lodge. I was way out in the country but, even after an hour and a quarter's walking, the abbey was still in sight through the mist. Fifteen minutes later it had gone, although it would probably have stayed longer in sight on a fine day. There was driftwood on the side of the bank, trees were still standing in the river thirty yards away, and I knew I was back in the peace of the country when a heron lifted from the waterside in front of me.

The width of the river here varies considerably, a hundred yards in parts, down to thirty-five in others, and making islands when it spreads itself. It is still flat country, with a lake and swans to the side.

After passing a stretch called The Nesses, I could make out Riccall, with its church tower and windmill, on the other side of the water. The river performs another U-turn at Wheel Hall, its nearest point to Riccall, and it was about here that the Norwegians, under their king, Harold Hardrada, moored with their 500 ships ready to conquer England in 1066. They marched through Riccall and on to York before retreating to defeat at the Battle of Stamford Bridge at the hands of King Harold, who was himself to die at Hastings in battle against William the Conqueror only weeks later. The remnants of the Norwegian army fled back to Riccall but were slaughtered as they went and, according to the Saxon Chronicle, '. . . few were left to carry back to Norway the dismal story.' In fact, only twenty-four of the 500 ships were needed for the return journey. The noise then must have been fearful. Now it was quiet as the river twisted its way past Kelfield. A partridge whirred across my path, which had moved sixty yards away from the river and provided room for parsnips to be grown in the narrow piece of land.

The path soon returned to the Ouse, where the tide had turned, enabling a car tyre to make its way lazily to Goole. The church bell at Cawood clanked out and I could make out the tower through the trees. I had almost reached one of the most historically interesting places on the entire walk, the village of Cawood, where kings and queens and archbishops used to stay and where Cardinal Wolsey was arrested for high treason.

Cawood Castle was a royal residence and a hunting seat, and King John went there in 1210 and 1212, followed by Henry III and his queen, Eleanor, who stayed there on a journey to Scotland. In 1251 Cawood witnessed a great scene of pageantry, when all the peers of the realm attended at Court, and Henry had a thousand knights in his train. Six hundred fat oxen, the gift of the archbishop, were slaughtered for the banquet.

There is a bridge across the river at Cawood, but there used to be a ferry across to the East Riding, and Ferry House was the principal inn, according to James Mountain in 1800. 'Indeed,' he wrote, 'it is the only one for the accommodation of travellers.' He said there were some good shops and a manufactory for hop-bagging. 'There is likewise the remains of a castle of great antiquity in which it is said Archbishop Mountain was born,' he wrote. 'He was likewise buried in the church of this place.' That archbishop must have had the shortest reign of any archbishop ever – or bishop, as well, come to that – for he died the same night after being enthroned at York. The bust of Archbishop Mountain, a poor farmer's son who was Bishop of Durham before becoming archbishop, is under an alabaster canopy in the church, showing him as if he were preaching, with his fingers in the leaves of a book.

The earliest account that James Mountain could find of Cawood when he was preparing his book *The History of Selby and Cawood* concerned the castle being built by King Athelstone about the year 920, and it was supposed that Athelstone gave the castle to the church in 930, so becoming the archbishops' palace.

The castle must have seen much magnificent pageantry in its days but little could have surpassed the day in 1470, during King Edward IV's reign, when George Neville was made archbishop

Entrance gate to the Bishop's Palace at Bishopthorpe, home of
the Archbishops of York today

and had a great feast to which were invited all the nobility, most of the Prime Clergy, and many of the gentry. The food that was needed was colossal and has been recorded and, as well as showing the scale of the feast, is also interesting for revealing the sort of food that was eaten five hundred years ago. There were:

300 quarters of wheat, 300 tuns of ale, 100 tuns of wine, one pipe of ypocrass, 104 oxen, 6 wild bulls, 1,000 muttons, 304 veals, 304 porks, 400 swans, 2,000 geese, 1,000 capons, 4,000 conyes, 204 bitterns, 400 heronshaws, 200 pheasants, 500 partridges, 400 woodcocks, 100 curlews, 1,000 egrittes, 500 and more stags, bucks and roes, 4,000 pasties of venison, cold, 2,000 pygges, 400 plovers, 100 dozen quales, 104 peacocks, 200 dozen fowls (called rees), 4,000 mallards and teales, 204 cranes, 2,000 chickens, 204 kids, 4,000 pidgeons, 300 parted dishes of jellies, 4,000 cold tarts baked, 3,000 cold custards baked, 1,500 hot pasties of venison, 2,000 hot custards, 608 pykes and breames, 12 purpoises and seals, spices, sugar'd delicacies and wafers plenty.

In 1530 Cawood had an illustrious visitor, Thomas Wolsey, Lord Chancellor, Bishop of Durham, Archbishop of York, once the chief minister of Henry VIII but now in disgrace. Arriving at Cawood, in his diocese of York, in the spring of 1530, he performed many charitable and popular acts, but in the following November he was arrested there, for high treason, and ordered to journey to London. But before Wolsey could reach his destination, he died, on 29 November 1530, and was interred in the chapel of the abbey of St Mary de Pratis at Leicester. It was said that he had taken poison.

Cawood was much larger in the days when the archbishops lived there, and the castle continued in all its splendour until the Civil War. For a time Cromwell's soldiers were garrisoned there until, on the order of Parliament, the castle, along with many other Yorkshire castles, was razed.

A footpath by the river at Cawood is aptly called Wolsey Walk and, after walking along Water Row, a path by houses alongside the river, I passed the Jolly Sailor, the Anchor and the Ferry Inn, before finding the site of Cawood Castle. The fifteenth-century

The Shambles at York

gatehouse is all that remains but I found another traveller who, like me, was prepared to stand and stare and imagine. 'This place is full of history,' he started and then told me all about Cawood and Wolsey. I had not the heart to stop him.

I made my way to the bridge, which was built in 1872 and where two men were sitting in a hut, whiling away the time. They told me the bridge had swung that morning for a barge heading upstream. 'We get 500-tonners,' I was told, 'Some heading for Rowntree's.' A ferry used to operate at this point but was replaced after a tragic incident in which the boat, carrying a carrier's covered waggon, was blown down the river and collided with a barge. People were thrown in the river and when a boatman saw what had happened he went to the rescue of the waggon driver and his wife. He could save only one of them, and the driver told the boatman to save his wife, Bessie, who lived many years to tell the tale. I asked at the hut how much of the 1872 bridge was still standing. 'This here's the genuine original except for what you're standing on,' I was told.

Again I had the problem of which side of the river to walk on. When the Ouse turns north at the junction with the River Wharfe only a mile away, the footpath to York, just about all the way, is on the west side. Yet there is no bridge, no way across – at least, none that I could see on the Ordnance Survey map. I asked at the hut and was told there was one way: I could swim it. I settled for the east side. At least that way I would stay with the Ouse, and there could well be a footpath on that side as well.

I crossed the bridge and walked alongside the water, past the mouth of the Wharfe where the Ouse turns sharp right, heading due north for York. The swing of the river gave me a new view of Cawood church across the fields to the right.

The Ouse is a lovely country river now, only about thirty yards wide in parts, but I soon had to leave it as the banking led away and over to the neighbouring road from Cawood to Stillingfleet and on to York. I did try for a while to stay with the river but the ground was thick and muddy and impossible, so I settled for the road and headed for Stillingfleet about a mile away, having to satisfy myself with the occasional early glimpse across the fields of the Ouse which was leaving me. To my right I could make out the cooling towers of the power stations at Drax and Eggborough, before reaching the Cross Keys, where the signpost pointed me to York and Naburn.

Stillingfleet Beck runs under the road below the church, and a tablet there has been erected by grateful parishioners to commemorate the great flood of 22–9 March 1947, when the depth of the water rose to twenty-one feet. The tablet also records the '. . . kindness shown by Capt. R. Boyle for the loan of the boat and Mr. A. Elcock and his crew'. The road leads between Moreby Hall and Bell Hall, two handsome old halls built in the early nineteenth and late seventeenth centuries respectively.

The river stays hidden for about forty minutes before appearing in full view just before Naburn Lock, which marks the end of the tidal part of the Ouse. There used to be a ferry at Naburn, too, until fairly recent times, commonly used by walkers and cyclists who could then make their way up to Bishopthorpe along the riverside path. For the time being I was still stuck to the right-hand bank as I wandered around the lock area with its barges. The first lock was opened in 1757 but a new one was needed and, after its opening in 1888, the first vessel through was the *Wild Rose*, carrying a cargo of 720 quarters of wheat, the largest shipment at the time to go through to York. Across the river is Acaster Melbis, with its cross-shaped fourteenth-century church where John Sharp, Archbishop of York in 1691, would often pray in peace and quiet. A barge lay sunk on the Selby side of the lock and, as I walked in the direction of York, a sign warned of another wreck, overgrown and below water, barely visible.

A series of stiles led me from the lock area towards the village of Naburn, about a mile away. The path ran out and I soon found myself at the back of Naburn Hall with its ruined chapel. Judging by all the barbed wire, I should not have been there, and I had to scramble over a gate to walk through Naburn, where a bridlepath leads to the river and the site of the old ferry. Ferry Farm Close and Ferry Cottage stood nearby as I pressed on past the Post Office, which is the size of a postage stamp yet which is also a Reading Room.

I soon returned to the river, where Naburn Marina was busy with boats, and then, quite unexpectedly, found my way across the river along the old railway line which ran from York to Doncaster. I made my way up the banking and onto the bridge, which is now concreted over and turned into a footpath. I had not known where I would cross the river, and the railway-line-cum-footpath came as a pleasant surprise. I returned to the riverside, happy in the knowledge now that, apart from a short detour at the palace at Bishopthorpe, I would be able to stay with the river right through to York. The walking was not as good as that I had known on the banking for so many miles on both the Trent and the Ouse but it was a path, just the same, and easy to follow.

There was a funny variety of gates and old back doors leading to boats moored on the river bank, one of them being just a door frame filled with barbed wire. I passed Birdcage Restaurant and Ferry Cottage as I kept to the path which led me to a huge cross close to the palace. A notice declares: 'On this spot there stood for centuries the Parish church of Saint Andrew Bishopthorpe. Rebuilt on another site AD 1899.' Dusk was approaching and this was an eerie spot, with its gravestones and remaining church wall. The ruin stands close to the palace wall, and Chantry Lane leads to the road running to York, only three miles away.

I put my nose inside the entrance to the palace grounds, through the beautiful entrance gate topped by a blue clock. A palace was originally built here in the thirteenth century by Archbishop Walter de Gray, but today's building is mostly eighteenth century with only some of the walling and a little chapel remaining of the original.

The view of the main entrance was impressive but I wished I could have seen it also from the other side of the river. I wished, too, that I could have gone inside and seen all the portraits hanging there of archbishops of York, including one of the unfortunate Scrope who took part in the rebellion against Henry IV and who was tried and convicted at Bishopthorpe. The Chief Justice of England refused to pronounce sentence against Scrope as a traitor, saying to Henry: 'Neither you, my Lord the King, nor any liegeman of yours in your name, can legally, according to the rights of the kingdom, adjudge any bishop to death.' Henry, presumably, could have passed sentence himself but instead ordered Sir William Fulthorpe, a lawyer, to condemn the Archbishop, who was executed close by. It was Archbishop Drummond who rebuilt the palace in the 1760s, and most of the stone used in the gateway was brought from the old palace at Cawood.

As I walked by the palace wall, I passed, on the other side of the

Back of the Guild Hall which stands close to the River Ouse in
York

road, the new Saint Andrew's Church, built here in 1899 after the site of the original on the banks of the Ouse had been abandoned due to bank erosion and flooding. At the end of the palace grounds a public footpath sign pointed me back to the river, a pleasant spot alongside the gently flowing water, now about forty yards wide. Ten minutes later I got another surprise when I came to steps leading up to the new road bridge built in 1976, Bishopthorpe Bridge, part of the A64 bypass between Leeds and Scarborough. I was desperately searching for my first sight of York Minster but I first had to reach the large, well-lit building, with a clocktower, of the chocolate manufacturers, Terry's, before I saw it.

It was dusk, hardly light enough for me to see to write, and there, floodlit and reaching up behind the trees, was the enchanting sight of the Minster. A shiver ran through me. I had walked from Westminster to York Minster but, while I felt a sense of achievement, it was the sight of one of England's loveliest buildings that thrilled me. A sign pointed me on to the city centre, twenty-five minutes away, and a road alongside the river led me past another chocolate connection, Rowntree Park on the left. Sturdy terraced houses stood across the river, and a few minutes later, sixteen days and 304 miles after leaving London, I walked away from the water, leaving the Ouse at Skeldergate Bridge. I walked past Clifford's Tower, along Castlegate and Coppergate, Colliergate and Petergate to the entrance of the Minster, a breathtaking building of great splendour.

York, a Roman city and a Saxon settlement, was invaded by the Vikings in 867. When northern England rebelled against the Norman Conquest, William the Conqueror crushed the uprising and built two wooden towers to guard the Ouse. One has totally vanished, leaving just the mound on which it stood, and the other was burned down the following century. Its stone replacement, Clifford's Tower, built in the thirteenth century, is still there, near the Castle Museum. York Minster was founded by Archbishop Walter de Gray early in the thirteenth century and stands on the site where Saxon and Norman churches have stood.

York is a fascinating place, crammed with history in much the way that York was crammed with people and buildings when it

The beautiful towers of York Minster . . . the end of the walk

was rebuilt by the Normans after a fire in 1069. York became a major fortress, and inside were forty churches, nine chapels, four monasteries, four friaries, sixteen hospitals and nine guildhalls. Four great bars (gates) still command the main roads to the city:

Micklegate, Bootham, Monk and Walmgate. York is a city for the walker, the wanderer, with so much to offer, so much to see. Narrow streets and old buildings are everywhere, and there is a feeling of warmth and permanence about the place.

But the end of my journey had to be in the Minster, where I said a prayer and thought of my sixteen days. When I went out into the streets, total darkness had descended. My journey was over.

OS MAP 105
Information Centres: Selby (0757 703263), York (0904 21756/7)

Bibliography

I referred to several books on the areas I covered, from London to
Yorkshire and found the following particularly useful:

Guides to the Waterways (Nicholson)
The Ordnance Survey Guide to the Waterways (Robert Nicholson,
  1983)
Grand Union Canal South (Waterways World, 1985)
Narrow Boat, L. T. C. Rolt (Eyre & Spottiswoode, 1971)
London North of the Thames, Arthur Mee (Hodder & Stoughton,
  1972)
Outer London, Simon Jenkins (Collins, 1981)
Hertfordshire, Arthur Mee (Hodder & Stoughton, 1965)
Uxbridge, Carolynne Hearmon (Hillingdon Borough Libraries,
  1982)
Half Hours with the Highwaymen, Charles G. Harper (Chapman
  & Hall, 1908)
Northamptonshire, Juliet Smith (Faber & Faber, 1968)
Northamptonshire, Tony Ireson (Robert Hale, 1974)
A History of Northamptonshire, R. L. Greenall (Phillimore, 1979)
Leicester Section Grand Union, River Soar and Erewash Canal
  (Waterways World, 1983)
Leicestershire and Rutland, Arthur Mee (Hodder & Stoughton,
  1967)
Leicestershire, W. G. Hoskins (Faber & Faber, 1970)
Portrait of the River Trent, Peter Lord (Robert Hale, 1972)
Nottinghamshire, Arthur Mee (Hodder & Stoughton, 1970)
Nottinghamshire, Cynthia Anne Shepperson (Dalesman Books,
  1976)
Discovering Nottinghamshire, Joan P. Alcock (Shire Publications,
  1978)
Nottinghamshire, Roy Christian (Batsford, 1974)
Slow Boat Through Pennine Waters, Frederic Doerflinger (Tan-
  dem, 1973)

A Tour Through the Whole Island of Great Britain, Daniel Defoe (Penguin, 1971 edition)

Journey to the Lake District by William Wilberforce (Oriel Press, 1983)

Diary of a Tour Through Great Britain in 1795, Rev. William MacRitchie (Elliot Stock, 1897)

Moritz's Travels in England, Carl Philipp Moritz (Humphrey Milford, 1924)

Lincolnshire, Henry Thorold and Jack Yates (Faber & Faber, 1965)

Lincolnshire, Walter Marsden (Batsford, 1977)

Yorkshire Villages, G. Bernard Wood (Robert Hale, 1980)

Yorkshire, Arthur Mee (Hodder & Stoughton, 1969)

Yorkshire Revealed, G. Douglas Bolton (Oliver & Boyd, 1955)

Portrait of the Yorkshire Ouse, Ivan E. Broadhead (Robert Hale, 1982)

Yorkshire: The West Riding, David Pill (Batsford, 1977)

Yorkshire: York and the East Riding, Nikolaus Pevsner (Penguin, 1972)

Yorkshire West Riding, Arthur Mee (Hodder & Stoughton, 1969)

History of Selby and Cawood, James Mountain (1800)

Yorkshire Legends and Traditions, Rev. Thomas Parkinson (Elliot Stock, 1888)

Yorkshire Ports and Harbours, Baron F. Duckham (Dalesman Books, 1967)

Canal Ports, J. Douglas Porteous (Academic Press, 1977)

History of the County of York, Thomas Allen (I. T. Hinton, 1831)

The Yorkshire Ouse, Baron F. Duckham (David & Charles, 1967)

History and Antiquities of Hemingbrough, Thomas Burton (Sampson Brothers, 1888)

History of Sherburn and Cawood, W. Wheater (Longmans, Green & Co., 1882)

History of the Isle of Axholme, Rev. W. B. Stonehouse (1839)

History of Collingham and Its Neighbourhood, E. G. Wake (Simpkin, Marshall & Co., 1869)

River Trent, J. H. Ingram (Cassell, 1955)

History of Winterton and Adjoining Villages, William Andrew (1836)

Index

Acaster Melbis, 191
Adam, Robert, 33
Adlingfleet, 163
Ad Pontem, 116
Aetheling, Edgar, 163
Aire and Calder Navigation, 171, 172, 181
Aire, River, 172, 175, 177
Albert Bridge, 23, 24
Albert Embankment, 21
Albert, Prince, 21, 66
Aldbury, 51
Alkborough, 162, 163
Althorpe, 148, 155, 157
Amcotts, 158, 159, 161
Apsley, 42
Apsley End, 48
Ashby de la Zouch, 97
Ashby St Ledgers, 70
Asselby, 175
Athelstone, King, 186
Attenborough, 102
Attenborough Nature Reserve, 101, 106
Avon, River, 71, 73
Axholme, Isle of, 148, 149
Aylesbury, 31, 42, 51, 61
Aylestone, 80, 81

Barlby, 185
Barmby on the Marsh, 175
Barnes, 28
Barrow upon Soar, 91, 92
Barton in Fabis, 101
Barton, Sir Thomas, 126
Barton upon Humber, 163

Batchworth, 38
Bathley, 127
Battersea, 23
Battersea Bridge, 24
Battersea Park, 23–4
Battersea Power Station, 22
Bawtry, 147
Beauvais, 126
Becket, Thomas, 50
Bedale, River, 165
Beeston Canal, 102, 108
Belgrave, 85, 86
Bell Hall (Yorkshire), 191
Belvoir, Vale of, 121
Benedict, Abbot, 181, 182
Berkhamsted, 11, 45, 46, 48, 50
Bevan, Benjamin, 62
Biddulph Moor, 99
Big Ben, 17
Birmingham, 57, 61, 69, 84
Birmingham and Warwick Junction Canal, 32
Birmingham Canal Navigation, 65
Birstall, 89
Bishopthorpe, 191, 192, 194
Blaby, 80
Black Prince, 50
Blacktoft, 164
Blacktoft Sands Nature Reserve, 164
Bleasby, 114
Bletchley, 57
Blisworth, 65
Blisworth Tunnel, 31, 65
Bole, 140
Bole Ferry, 140

Bonaparte, Napoleon, 108
Boots the Chemist, 102, 106
Boston, 137, 138
Boston Manor, 33
Bosworth, Battle of, 70
Boxmoor, 11, 42, 43, 45, 48
Bradford, 175
Braunston, 31, 38, 41, 43, 55, 61, 69
Brentford, 11, 13, 29, 31, 32, 33, 37, 43, 61
Brent, River, 33, 34
Bridgewater, Duke of, 51
Brunels, 24, 34
Buckby Locks, 12, 13
Buckingham, 31, 62
Bugbrooke, 66
Bulbourne, 51
Bull's Bridge, 34
Burghley House, 66
Burgh, Sir Thomas, 143
Burlington, Earl of, 28
Burringham, 155
Burton Joyce, 108, 109
Burton Power Station, 138, 140
Burton Round, 133, 140
Burton, Thomas, 177, 179
Burton upon Stather, 151, 161, 162, 163
Burton upon Trent, 99, 144
Burton, William, 178
Byron, Lord, 101, 123

Caesar, Julius, 32
Caldon Canal, 65
Canterbury, Archbishop of, 18, 20
Canterbury Cathedral, 50
Canute, King, 32, 143
Carlton, 129
Caroline, Queen, 50
Cassiobury Park, 41
Castleford, 177
Castlethorpe, 63

Catesby, Robert, 70
Catesby, William, 70
Catterick, 165
Cawood, 186, 188, 190, 192
Caythorpe, 112
Charles I, King, 35, 78, 80, 111, 122, 148, 167, 168
Charles II, King, 21, 23, 38, 42
Charnwood Forest, 92
Cheddington, 51
Chelsea Bridge, 22
Chelsea Embankment, 22
Chelsea Flower Show, 23
Chelsea Royal Hospital, 23
Chess, River, 39
Chesterfield Canal, 146, 147
Cheyne Walk, 24
Chiswick Bridge, 28, 29
Chiswick Eyot, 28
Chiswick House, 28
Churchill, Winston, 18
Civil War, 11, 32, 35, 41, 78, 80, 102, 108, 111, 122, 123, 138, 188
Clarendon, Earl of, 41
Clarke, William, 105
Cliffe, 176, 177
Clifton, 102
Clifton, Sir Robert Juckes, 103
Clore Gallery, 21
Cobbett, William, 121
Codbeck, River, 165
Collingham, 128
Colne, River, 37, 38, 39
Colwick, 106, 109, 117, 133
Colwick Park, 106
Cosgrove, 61, 62, 63
Cossington, 90
Cottam Power Station, 128, 134, 137, 138, 140
Countesthorpe, 78
Cowley Peachey, 34
Cowper, William, 48, 49
Cowroast, 50, 53

Cranfleet Cut, 99, 108
Crick, 71
Cromwell, Oliver, 18, 35, 102, 122, 188
Cromwell village, 128
Croxley Green, 39

Defoe, Daniel, 121, 163, 181
Denbigh, Countess of, 78
Denbigh, Earl of, 57
Denbigh Hall, 57
Denham, 37
Denley, John, 35
Derrythorpe, 155
Derwent Mouth, 101
Derwent, River, 175, 176
Devon, River, 119
Doddridge, Dr Philip, 77
Domesday Book, 43, 134, 138, 149, 157, 161, 176
Doncaster, 139
Don, River, 148, 164, 168
Dover Beck, 112
Drax, 177
Drax Power Station, 175, 177, 190
Drayton, Michael, 167
Duckham, Baron F., 181
Dudley Tunnel, 65
Dunham on Trent, 125, 128, 131, 133
Dunstable, 54
Dutch River, 168, 172

East Bridgford, 111
East Butterwick, 154
East Ferry, 151, 152
East Retford, 147
East Stockwith, 146
East Stoke, 114, 115
Edgecote, 63
Edward the Confessor, 18
Edward I, King, 130, 159, 163
Edward II, King, 158, 162

Edward III, King, 18, 149
Edward IV, King, 63, 186
Eggborough Power Station, 177, 190
Eleanor of Castile, 130
Eleanor, Queen, 186
Eliot, George, 144
Elizabeth I, Queen, 33, 151, 157
Elizabeth II, Queen, 69
Epworth, 148, 149, 167
Erewash Canal, 32, 99, 100
Erewash, River, 102
Etruria, 65

Farndon, 13, 117, 119
Farnham, John, 92
Fawkes, Guy, 70
Fenny Stratford, 57, 62
Festival of Britain, 23
Fiennes, Celia, 121
Fiskerton, 115, 116
Fledborough, 130, 131
Flintham Wood, 114
Flixborough, 159, 161
Flynn, Errol, 105
Folly Drain, 148
Fossdyke and Witham Navigations, 65, 137, 138
Fosse Way, 89, 115, 116, 121
Foxe's *Book of Martyrs*, 35, 43
Foxton, 75
Foxton Locks, 14, 71, 73, 75
Fulham FC, 26
Fulham Palace, 26
Fulthorpe, Sir William, 192
Furnivall, Dr Frederick James, 28

Gade, River, 39, 42, 43
Gainsborough, 11, 99, 125, 133, 134, 140–5, 147, 151, 152, 162
Garnett, Father, 101
Garthorpe, 161
Gate Burton, 139

Gayton, 66
George I, King, 32, 50
George II, King, 50
George III, King, 22, 66
George V, King, 18, 157
George VI, King, 69
Girton, 129
Gladstone, William, 123
Glen Parva, 80
Goole, 165, 167, 168, 171, 172, 173, 176, 177, 181, 186
Grafton Regis, 63
Grand Junction Canal, 31, 32, 37, 38, 54, 65
Grand Union Canal, 11, 12, 13, 14, 31–97, 128
Grantham, 122
Great Brickhill, 55
Great Linford, 59
Great North Road, 121, 125, 126
Great Ouse, River, 31, 61, 62, 63
Great West Road, 33
Great Woolstone, 59
Green, George, 106
Greene, Graham, 48
Gresham, Sir Thomas, 33
Grimsby, 148
Grove Park, 41
Gumley, 77
Gunhouse, 147
Gunness, 158, 161
Gunpowder Plot, 70, 101
Gunthorpe (nr. Gainsborough), 145, 148, 151
Gunthorpe (nr. Nottingham), 109, 111

Hacker: Col. Francis, Rowland and Thomas, 111
Hadrian, Emperor, 89, 139
Halford, Andrew, 78
Halford, Sir Richard, 78
Hammersmith, 26, 28

Hammersmith Bridge, 26
Hammersmith Reach, 28
Hanover, 32, 50
Hanwell, 33, 34
Harby, 130
Harefield, 38
Harold Hardrada, King, 185
Harold, King, 139, 163, 185
Harper, Charles, 45
Hastings, 139, 185
Haxey, 149
Hazelford Ferry, 113, 114
Hearmon, Carolynne, 35
Heathrow, 26, 28, 33
Hemel Hempstead, 42, 43, 45, 47, 54
Hemingbrough, 176, 177
Henry II, King, 149
Henry III, King, 90, 159, 186
Henry IV, King, 192
Henry VI, King, 128
Henry VII, King, 114, 115
Henry VIII, King, 24, 29, 32, 37, 143, 188
Herbert, A. P., 28
Hertford, 46
High Marnham, 130
Holme, 13, 125, 126, 127
Hood Game, 149
Hook, 172, 173
Horncastle, 137
Horseley Iron Works, 33
Household Cavalry, 69
Hoveringham, 112
Howard, Catherine, 29
Howden, 181
Howden Dyke, 173
Hull *see* Kingston upon Hull
Humber Bridge, 163
Humber, River, 99, 127, 145, 147, 148, 157, 162, 163, 164, 167, 182
Hungerford, 45, 46
Huntington, William, 147

Hunton Bridge, 41, 42
Husbands Bosworth, 73

Idle, River, 146, 147, 148
Incent, John, 48
Ireton, John and Henry, 102
Ivinghoe, 53

James VI and I, King, 29
James, Jack, 65
Jermyn family, 138
John, King, 122, 130, 186
John Player, 106
Johnston, Edward, 28
Julian's Bower, 163

Keadby, 145, 157, 158
Kegworth, 95, 101
Kelfield (Humberside), 145, 152, 154
Kelfield (North Yorkshire), 185
Kennington Oval, 22
Kew Gardens, 29
Kibworth Beauchamp, 77
Kibworth Harcourt, 77
Kilsby tunnel, 57
Kimberley, Relief of, 23
King's Causeway, 167
Kings Langley, 42
Kings Road, Chelsea, 25
King's Royal Rifles, 159
Kingston upon Hull, 144, 151, 154, 157, 161, 163, 165, 167, 172, 173
Kinnaird's Castle, 149
Kirk, Sir John, 95
Kitchener, Lord, 90
Knaith Hall, 139
Kneeton, 111, 112
Knights Templar, 90

Lake District, 133
Lambeth Bridge, 18

Lambeth Palace, 20, 21
Lancaster, Osbert, 23
Laneham, 13, 134, 135
Langley Mill, 99
Laughton Common, 148
Lawrence, D. H., 99
Lea, 139
Leeds, 175
Leicester, 11, 13, 14, 31, 69, 73, 77, 78, 80–6, 93, 188
Leicester and Swannington Railway, 83
Leicester Journal, 140
Leicester Navigation, 32, 83, 93
Leicestershire and Northamptonshire Union Canal, 31–2, 80
Leighton Buzzard, 11, 54, 55, 62
Leland, John, 37
Lincoln, 84, 137, 138, 139, 158
Lincoln, Bishop of, 122, 123
Lincoln Cathedral, 122, 137
Linslade, 54
Linton upon Ouse, 165
Liscombe House, 55
Littleborough, 139
Little Reedness, 167
Little Tring, 51
Little Woolstone, 59
LMS (London, Midland & Scottish Railway), 70
London, 17–29, 45, 50, 51, 62, 63, 66, 70, 80, 83, 84, 130, 144, 149, 188
London & North-Western Railway, 61
London, Bishops of, 26
London Bridge, 17
Long·Buckby, 69
Long Drax, 176
Long Eaton, 97, 99, 137
Loughborough, 84, 89, 91, 92, 93, 121

Loughborough Navigation, 32,
92, 93
Low Melwood, 149
Luddington (Humberside), 161

Macaulay, Thomas, 90
Malham Moor, 175
Manchester, 84, 172
Margaret, Princess, 106
Market Bosworth, 73
Market Harborough, 75
Marnham Power Station, 128, 130
Marsworth, 51
Marton, 138
Mary, Queen of Scots, 161
Massachusetts, 138
Meadows, The (Nottingham),
103
Mee, Arthur, 59, 77, 123, 135
Melton Mowbray, 121
Mercia, King of, 66
Millbank Prison, 21
Milton Keynes, 59, 63
Milton Malsar, 66
Misbourne, River, 37
Misterton, 146, 147
Monmouth, Duke of, 38
Moore, Henry, 21
Moor Park, 38
Moreby Hall, 191
More, Thomas, 24
Morton, 146
Mother Drain, 147
Motorways: M1, 70; M4, 33; M6,
66; M25, 42; M62, 172, 173
Mountain, James, 181, 186
Mountsorrel, 90, 121
Mowbray, Lady, 149

Naburn, 191
Naburn Lock, 165, 191
Naseby, Battle of, 78, 80
National Army Museum, 23

National Trust, 33
Nene, River, 67
Nether Heyford, 66
Newark-on-Trent, 11, 13, 99, 114,
117–26, 133, 138, 144
New Bradwell, 61
Newcastle, Duke of, 125
Newhay, 176
Newlands Park, 59
Newton Harcourt, 78
Noel-Buxton, Lord, 17
Normanby, 161
Normanby Hall, 162
Normanton on Soar, 93
Northampton, 31, 63, 66, 84
Northampton Mercury, 37, 62
Northchurch, 50
North Clifton, 130, 131
North Collingham, 128
North Kilworth, 73
North Muskham, 125, 126, 127
Northumberland, Earl of, 29
Norton Junction, 31, 69
Nottingham, 11, 12, 13, 31, 69, 84,
99, 103–9, 116, 133, 138, 167
Nottingham Forest FC, 103
Nottinghamshire CCC, 103
Notts County FC, 103

Old Goole, 168, 171
Old Linslade, 55
Old Stratford, 62
Ossington, Viscountess, 123
Osterley House, 33
Ousefleet, 165
Ouse, River, 11, 12, 13, 99, 109,
157–96
Ouzel, River, 55, 57
Owlet Plantation, 148
Owston Ferry, 145, 149, 151, 152,
154, 162, 167
Oxford, 35, 66, 84
Oxford Canal, 31

Paddington, 31, 34
Pankhurst, Sylvia, 18, 24
Parliament, Houses of, 17, 18
Parr, Katherine, 143
Paunceforte, Julian, 115
Paupers' Drain, 148, 159
Peace Pagodas, 24, 59
Pevsner, Nikolaus, 123
Pilcher, Percy, 71
Piper, John, 23
Pitstone, 51
Plague, 13, 127
Pocklington, 167
Porteous, J. Douglas, 181
Potter Hill, 128
Preston, 172
Putney, 28
Putney Bridge, 26

Quorndon, 91, 92

Radcliffe on Trent, 108, 133
de Radcliffe, Stephen, 108
Radford, 130
Ragged Schools, 95
Raleigh bicycles, 106
Raleigh, Sir Walter, 161
Ranelagh Gardens, 22
Ratcliffe on Soar, 97
Ratcliffe Power Station, 95, 99, 101
Reedness, 165, 167
Reeth, 165
Regent's Canal, 32
Riccall, 185
Richard I, King, 18, 178
Richard II, King, 18, 42, 123
Richard III, King, 70, 115, 143
Richmond, 165
Rickmansworth, 37, 39
Ripon, Marquess of, 24
Robin Hood, 105
Rolleston, 116
Rolt, L. T. C., 81, 85, 92

Rothersthorpe, 66
Rothley, 90
Rowntree's of York, 190, 194
Roxby, 161
Rupert, Prince, 122
Rusholme Grange, 175

Saddington Tunnel, 77
St Germanus, 181, 182
St Werburgh, 66
Salisbury, 126, 182
Saltmarshe, 167
Sawley Cut, 97, 108
Saxon Chronicle, 185
Scott, Nan, 13, 126, 127
Scott, Sir Walter, 123
Scunthorpe, 148, 154, 155, 157, 158, 161
Sebert, King of East Saxons, 18
Sedgemoor, Battle of, 38
Segelocum, 139
Selby, 11, 165, 173, 176, 177, 179, 181, 182, 183, 185
Selby Abbey, 13, 177, 179, 181, 182, 183
Selby Canal, 171, 181
Sence, River, 77
Severn, River, 99, 145
Shaftesbury, Lord, 95
Sheffield and South Yorkshire Navigation, 158
Sheffield, Sir Charles, 162
Sheffield, Sir John, 165
Sheffield, Sir Robert, 162
Shelford, 108
Sherwood Forest, 105
Sileby, 90
Simpson, 57
Slapton, 53
Slough, 31, 34
Smeeton Westerby, 77
Smith, Charles John, 151
Snaith, 168

Sneinton, 106
Snooks, Robert, 43, 45, 46, 47, 48
Soar, River, 11, 78, 80, 81, 85, 89,
  90, 92, 95, 100
Soulbury, 55
South African (Boer) War, 23, 173
South Butterwick Drain, 154
South Clifton, 129, 130, 131
South Collingham, 128
South Harefield, 37
South Muskham, 125
Southwark, 45
South Wigston, 78
Stainforth and Keadby Canal, 158
Stamford, 138
Stamford Bridge, Battle of, 185
Staythorpe, 114
Staythorpe Power Station, 117
Stephenson, George, 84
Stephenson, Robert, 83
Stillingfleet, 190
Stirling, James, 21
Stoke Bardolph, 108, 109
Stoke Bruerne, 61, 63, 65, 66
Stoke Hall, 116
Stoke Hammond, 55
Stoke Park, 65
Stonehouse, Revd W. B., 145, 151,
  152, 154, 155, 157, 159, 161
Stovin, George, 159
Strand-on-the-Green, 28
Stratford and Buckingham Canal, 62
Sturton le Steeple, 139
Surrey CCC, 22
Susworth, 152
Sutton Bonnington, 95
Swale, River, 165
Swannington, 83
Sweetapple, William, 131
Sweyne, King of Denmark, 143,
  163, 167
Swinefleet, 167, 171
Syon House, 29, 32

Syston, 89, 90

Tarbotton, Marriott Ogle, 105
Tate Gallery, 21
Tattershall Castle, 128, 138
Terry's of York, 194
Thames, River, 17–29, 31, 32, 34,
  37, 51, 54, 91, 99
Thonock, 143
Thorne, 144
Thorney, 130
Thornton, 62
Thrumpton, 100
Thrumpton Hall, 101
Thurgarton Priory, 115, 116
Thurmaston, 89
'Tom Puddings', 172
Torksey, 137, 138, 139, 143
Tove, River, 63
Trent and Mersey Canal, 100
Trent Bridge, 103, 105
Trent Falls, 164, 168
Trent Hills, 114
Trentlock, 12, 13, 90, 97, 99, 101
Trent Ness, 164
Trent Port, 138
Trent, River, 11, 12, 13, 14, 69, 71,
  78, 93, 99–165, 173, 192
Tring, 51, 61
Turner, J. M. W., 21, 24
Turnham Hall, 177, 178, 179
Turpin, Dick, 126

Ulverston, 181
Upper Heyford, 66
Ure, River, 165
Uttoxeter, 65
Uxbridge, 11, 33, 35, 37, 129

Vauxhall Bridge, 21, 22
Vauxhall Gardens, 21, 22, 23
Vermuyden, Cornelius, 148, 155,
  168
Victoria, Queen, 21, 34, 66, 173

Wake, E. G., 126, 128
Wakefield, 172
Walkeringham, 146, 147
Walkerith, 146
Walker, Sir Emery, 28
Wanlip, 89
Warwick and Birmingham Canal, 32
Warwick and Napton Canal, 32, 34
Washington, John, 183
Wash, The, 137
Watford, 39, 41, 42
Watford Gap, 70
Watling Street, 57, 66
Weedon Bec, 66, 67
Welford, 71, 73
Wellington, Duke of, 23, 108
Wendover, 31, 51
Wesley, John, 121, 148, 167
West Burton, 140
West Butterwick, 151, 154, 155
Westcomb, Lucy Emerton, 101
West Drayton, 34
West Ferry, 145, 151, 152
West Kinnard Ferry, 151
Westminster, 11, 13, 17, 28, 42, 70, 194
Westminster Abbey, 13, 18, 90
Westminster Bridge, 17, 18, 21
Westminster Hall, 18
West Stockwith, 13, 146, 147, 148, 157
Westwoodside, 149
Wharfe, River, 190
Whipsnade Lion, 53
Whitgift, 165, 167
Wigston, 78
Wilberforce, William, 133, 140
Wildsworth, 149
Wilford, 103
Willen, 59
William II, King, 18

William III, King, 133
William IV, King, 43
William the Conqueror, 50, 164, 182, 185, 194
Windsor, 32
Windsor, Duke of, 111
Winkwell, 48
Winthorpe, 119, 125, 126
Wintringham, 161
Wistow, 78
Witham, River, 137
Wolsey, Cardinal, 38, 39, 85, 186, 188
Wolverton, 31, 61, 62
Wolvey, 80
Wood, G. Bernard, 176, 177
Wood, George Derwent, 24
Woodville, Elizabeth, 63
Worcester, 122
Worcester, Battle of, 23
Worksop, 147
World War I, 62
World War II, 33, 114, 161
Woughton on the Green, 58
Wreake, River, 90
Wren, Sir Christopher, 23
Wright Brothers, 71

Yardley Gobion, 63
Yelvertoft, 71
Yokefleet, 165
York, 11, 12, 14, 84, 126, 139, 163, 165, 167, 185, 186, 190, 191, 192, 194, 195, 196
York, Archbishops of:
    Drummond, 192; de Gray, Walter, 192, 194; Mountain, 186; Neville, George, 186; Paulinus, 114; Scrope, 192; Sharp, John, 191
York Minster, 13, 177, 194, 196

Zouch, 95